Health

——— AND ———

Numbers

Basic
Biostatistical
Methods

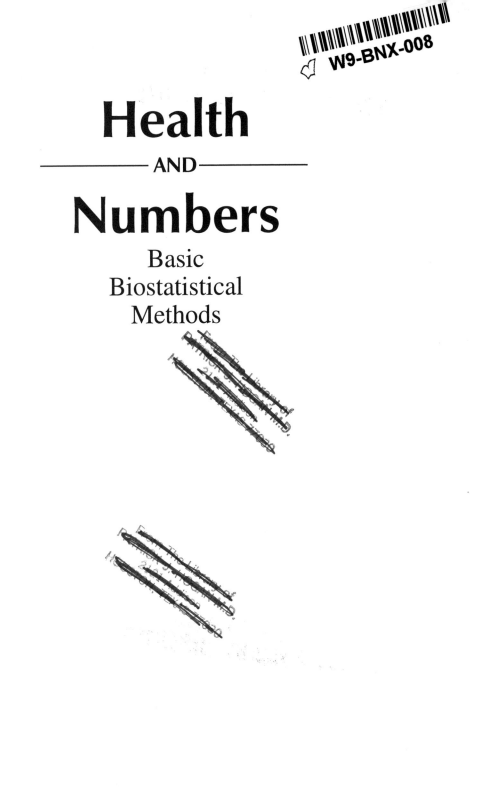

Health

———— AND ————

Numbers

Basic
Biostatistical
Methods

Chap T. Le
James R. Boen

Professors of Biostatistics
School of Public Health
University of Minnesota

 **WILEY-LISS**

A JOHN WILEY & SONS, INC., PUBLICATION
New York • Chichester • Brisbane • Toronto • Singapore

Address all Inquiries to the Publisher
Wiley-Liss, Inc., 605 Third Avenue, New York, NY 10158-0012

Copyright © 1995 Wiley-Liss, Inc.

Printed in the United States of America.

Library of Congress Cataloging-in-Publication Data

Le, Chap T., 1948–
    Health and numbers : basic biostatistical methods / Chap T. Le,
James R. Boen.
      p. cm.
    Includes bibliographical references and index.
    ISBN 0-471-01248-3
    1. Medical statistics. 2. Biometry. 3. Statistics. 4. Medicine-
-Research—Statistical methods. I. Boen, James R., 1932–
II. Title.
    [DNLM: 1. Biometry—methods. WA 950 L433h 1994]
RA409.L42 1994
362.1'015195—dc20
DNLM/DLC
for Library of Congress   94-25553
                                   CIP

The text of this book is printed on acid-free paper.

10 9 8 7 6 5 4 3 2 1

Printed and bound by Malloy Lithographing, Inc.

To my wife, Minha, and my two daughters,
Mina and Jenna, with love.

—Chap T. Le

Contents

Preface

We have taught introductory biostatistics courses for many years to students from various human health disciplines, and now we have decided it's time to write our own text. As teachers, we are very familiar with students who enter such courses with an enormous sense of dread, based on a combination of low self-confidence in their mathematical ability and a perception that they would never need to use statistics in their future work, even if they did learn the subject. Those students who enter health fields that emphasize caring and human contact are especially apt to view statistics as the antithesis of interpersonal warmth. We are sympathetic to the plight of students taking statistics courses simply because it is a requirement and consequently have tried to write a friendly text. By friendly we mean *a fairly thin book* and *emphasizing a few basic principles* rather than a jumble of isolated facts. The perception that statistics is just a bunch of formulas and long columns of numbers is the main misunderstanding of the field. Statistics is a way of thinking, thinking about ways to gather and analyze data. The formulas are tools of statistics, just like stethoscopes are tools for doctors, and wrenches are tools for auto mechanics. The first thing a good statistician does when faced with data is to learn how the data were collected; the last thing is to apply formulas and do calculations. It's amazing how many times statistical formulas are misused by a well-intentioned researcher simply because the data weren't collected correctly. Start to pay attention to the number of times the media will announce that studies have been done that lead to such-and-such a conclusion, without telling you how the study was done or the data collected. Most of the time the reporters don't even seem to know that the methods of the study or the method of data collection are crucial to the study's validity; their job is to seek out newsworthy findings and dramatize the implications.

Almost all of the formal statistical procedures performed fall into two categories: testing and estimation. This book covers the most commonly used, elementary procedures in both categories: the Z-test, the t-test, the chi-square test, point estimation, confidence interval estimation, and regression/correlation. These are the nuts and bolts of elementary applied statistics. We also throw in some basic ideas on standardization of rates and graphical techniques that are easy to learn and incredibly useful. Exercises are added as aids to learning how to use statistical procedures. Learning to use statistics takes practice. It's not an easy subject, but it's worth learning. Let us know how we can improve the book.

The book is divided into seven chapters. Chapters 1 and 2 introduce descriptive statistics, the art of organizing, summarizing, and presenting data; Chapter 1 is concerned with discrete outcomes and Chapter 2 with continuous measurements. Chapter 3 briefly introduces the concept of probability and some frequently used probability models, a bridge between descriptive an inferential statistics. Inferential statisics, the science of generalizing from observed samples to target populations, is covered in Chapters 4, 5, and 6. Chapter 4 is concerned with the subject of confidence estimation, and Chapters 5 and 6 are on statistical tests of significance; Chapter 5 introduces the basic philosophy and foundation and Chapter 6 the methods. Finally, Chapter 7 introduces a few other selected topics; presentations are more brief. The book is designed for a 3-credit semester course or a 4-credit quarter course; for shorter courses, the instructor may decide not to cover Chapter 7 and some parts of Chapter 6.

The authors would like to express their appreciation to teaching assistants who helped to select some of the exercises and especially to Deborah Sampson for her usual hard work and patience in the preparation of the manuscript.

Introduction

Recently we met Carmen, a middle-aged woman. During social chit-chat, she told us that she is a nurse on a cancer ward. When we announced that we are statistics professors, she said she was planning on getting her master's degree in nursing and that we might be interested in some advice she heard from other nurses who get master's degrees. "If you get only one chance for a pass–fail option, use it on the stat course," they counseled. They made it clear that the statistics course required of graduate students in nursing is really hard and that simply surviving it is the only practical goal.

We had no problem believing Carmen, nor did we question the sincerity of the students' recommendation. Many times that sentiment is expressed: Statistics courses are experiences to be endured, and even honest people are tempted to cheat in order to pass. Survival skills are in order: "Do anything, including neglecting the other courses for one quarter, to get it behind you. Use flash cards, buy other books, make friends with nerds if you must, but get through the requirement."

These feelings are expressed in many ways and in many different settings by those who have taken required statistics courses. In the social setting of a party, the revelation that we teach statistics usually provokes such comments from a course survivor or one who is dreading the experience. Indeed, the required statistics course may well be described as a hostage situation, with the students as prisoners and the teacher as torturer.

The drama of the required statistics course is re-enacted year after year throughout the country. Students from a wide variety of helping professions take statistics courses as they prepare for a career. This book is written especially for those who feel inadequate with statistics in any form, but are forced by journal editors, government funding agencies, or degree requirements to deal with it. If you are one of those people, this book is for you.

Statistics is a hard subject; many people, however, realize that it is essential, not only in science and government but even in the human services fields. As the federal budget problems continue, both the Pentagon and the Department of Health and Human Services cite statistics to help their cases. Applicants for increased government funding for worthy causes can no longer claim only their good intentions as the sole reason for a bigger chunk of the pie. They have to present numbers to show that funding their cause is a better bargain "for the people" than funding a competing worthy cause. This creates a demand for an objective way of thinking, and an art of communication, called *statistics*.

The "way of thinking" called statistics has become important to all professionals who are not only scientific or business-like, but are caring people who want to help to make the world a better place. Such people have big hearts, high ideals, and a concern for humanity. They are good people who work hard and expect only a modest income. They become teachers or nurses or clergy or social workers, the kind of people inclined to join the Peace Corps. They even do volunteer work in a wide variety of human services. Such people are the special audience of this book. Due to funding pressures and self-imposed requirements that their field be more research-based, those who seek higher degrees are forced to take statistics courses, read journal articles, and write reports using numbers and statistical formulas. They and their administrators must defend their work by counting and measuring and defend against those who attack their counting and measuring methodology.

This book is written for helpers who are *forced* to deal with the world of statistics, but feel inadequate because of low mathematics ability, lack of self-confidence, or the perception that they will never need to use statistical concepts and techniques in their future work.

WHAT IS STATISTICS?

There are several popular public definitions and perceptions of statistics. We see "vital statistics" in the newspaper, announcements of life events such as births, marriages, and deaths. Motorists are warned to drive carefully, to avoid "becoming a statistic." The public use of the word is widely varied, most often indicating lists of numbers, or data. We have also heard people use the word *data* to describe a verbal report, a believable anecdote. Statisticians define statistics to be summarizing numbers, like averages; e.g., the average age of all mothers on AFDC is a statistic, because it is a (partial) summary of a list of numbers too long and too varied to describe individually. The average is a useful partial summary and thus a statistic.

For this book and its readers, we don't emphasize the definition of "statistics as things," but offer instead an active concept of "doing statistics." The doing of statistics is a way of thinking about numbers, with emphasis on relating their interpretation and meaning to the manner in which they are collected. The relation of method of collection to technique of analysis, by the way, is absolutely central to the understanding of statistical thought. Anyone who claims facility with statistical formulas but is oblivious to the method of collection of the data to be analyzed must not portray himself or herself as a statistician. Failing to see the important relationship of collection to analysis is the root cause of mindless throwing of statistical formulas at data.

Our working definition of statistics, as an activity, is that it is a way of thinking about data, about its collection, its analysis, and its presentation to an audience. Formulas are only a part of that thinking, simply tools of the trade.

To illustrate statistics as a way of thinking, we notice that our local radio station has a practice of conducting radio polls and announcing their results during the morning rush hour. One question was whether an old baseball player (who was personally popular here and had a high batting average) should be recruited to play for the Minnesota Twins again. Of the people who called in their answer, 74% said "yes" and 26% said "no." As people who are well-groomed into statistical thinking, we wondered whether the 74% meant that 74% of the people in the city want him back, or 74% of baseball fans want him back, or what? In trying to figure whether the 74% meant anything, we listened to the method by which the data were collected. The announcer said that only people who have touch-tone phones can vote because only such phones will work for their voting system. So now the poll is limited to people with touch-tone phones who aren't too busy at 8:30 a.m. Of those who have the phones and the time, what kind of people care enough to vote? Thinking through this maze of selections is doing statistics. All this kind of thinking must be done before applying any statistical formulas makes sense. There are many people who don't know statistical formulas, but naturally think statistically. Someone who throws statistical formulas at data but doesn't appreciate the subtleties of data collection is dangerous.

WHAT CAN STATISTICS DO?

There is a broad spectrum of beliefs among nonstatisticians about what the field of statistics can actually do. At one extreme are the cynical disbelievers who think the only contribution of statisticians was the discovery that, at cocktail parties, two percent of the people eat ninety percent of the nuts. At the other extreme are the blind-faith believers who envision statistics as a crime lab going over the murder room with a fine-tooth comb. They also think of statisticians as archaeologists digging gingerly into mounds of data, extracting from them every bit of truth.

WHY IS STATISTICS SO HARD TO LEARN?

Statistical formulas commonly result in numbers that have no intuitive meaning. They do not relate to any ordinary experience. At the end of the calculations for a t-test, the number 1 is considered very small, but the number 3 is very large. This defies intuition. The person performing a *t*-test looks in *t*-tables at the back of a statistics book to find that the probability associated with 1 is .32. Next, the reader of the table is told that .32 is too large to be called significant. It is *not* significant. You might well think that .32 is not a significant number in some ordinary sense, but because it is too small. The *t*-table tells you it's *too large* to be significant. If, after calculating the *t*-test, you get the "large" number 3 and then look in the *t*-table to find the probability associated with it is .005, you are told that .005 is small enough to be highly significant. What is this, anyway? How does a 1 get you a .32 while a 3 leads to a .005? What's going on when little numbers like .005 are highly significant? Any logical person thinks big numbers are significant, not small ones. This turning upside down, these brand new meanings assigned to small numbers and large ones are a major adjustment for the student of statistics. Anyone who gets confused converting inches to centimeters or Celsius to Fahrenheit is in for hard work learning to use statistical formulas.

One class of volunteers who want to learn statistics are the graduate students seeking a master's or doctoral degree in biostatistics, a specialization of statistics. These students study full-time for at least a year before starting to get an integrated notion of statistics, and many of them come to our program already having earned a bachelor's degree in mathematics.

The nonvolunteers have an even worse time. The ones we see are the graduate students in fields such as nursing or environmental sanitation who are required to take one or two statistics courses for their masters or doctoral degrees. When we teach them, we do what we can to minimize their frustration with the material. Some of them seem to learn the formulas fairly easily, but few of them can get a good handle on the concepts. A good number of those students spend an inordinate amount of time on the course, neglecting their other courses, in order to pass statistics and get beyond the hurdle. Several students try hard to find a statistics course taught by an easy grader. They dread the statistics requirement.

REASONS FOR THE DIFFICULTY

There are two kinds of reasons for statistics being a difficult subject to learn: intellectual and emotional.

Intellectual Reasons

There are concepts in statistics that are difficult. For one thing, although statistics is definitely not the same as mathematics, it *uses* mathematics. Anyone who has trouble with arithmetic and algebra has trouble with statistics, and there are lots of people with excellent communication skills (reading, writing, speaking) who have trouble with arithmetic and algebra. The Graduate Record Exam, as well as other standard examinations such as the Scholastic Aptitude Test taken by high school students, has the separate categories of verbal and quantitative; they are different. To attain the masters degree of knowledge of statistics requires a good knowledge of calculus and some facility with a topic called *matrices* or *linear algebra*. Doctoral degree knowledge of statistics requires, in addition, advanced calculus, and something called *measure theory* as a basis for understanding probability. In other words, getting deeply into statistical theory requires a knowledge of mathematics attained by very few people. The mathematics of advanced statistics is approximately the level of the mathematics of theoretical physics.

In addition to the mathematics of statistics, there are other difficult concepts. The main difficulty is in visualizing the distribution of numbers you cannot see. Many students have a hard time getting straight the sampling distribution of the sample means, when the end point of the study is only one sample mean. That is, they have to be able to visualize infinitely many numbers, of which they see only one, and think about what the other numbers might have been.

Emotional Reasons

There are aspects of statistics other than it being intellectually difficult that are barriers to learning. For one thing, statistics does not benefit from a glamorous image that motivates students to persist through tedious and frustrating lessons. A pre-medical or pre-law student is commonly sustained through long discouraging times in school by dreams of wealth, high social status, and heroism in the not-too-distant future. However, there are no TV dramas with a good-looking statistician playing the lead, and few mothers' chests swell with pride as they introduce their son or daughter as "the statistician."

The public images of statisticians leave much to be desired as sources of recruitment. *How To Lie With Statistics* (by Darrell Huff) is the statistics book whose title is most often quoted by non-statisticians. One image of statisticians is that of sports nuts who are fascinated by numerical trivia of games. Another image is of the librarian, the solemn keeper of dry details, of "more than you'd ever want to know about ...." Yet another is the role of the manipulator of numbers, the crook who can make numbers say anything he or she wants them to. Presidential campaigns in which both incumbent and challenger cite "statistics" showing why they should be elected don't give statistics a good name.

Have you ever heard of a child who answered the question, "What are you going to be when you grow up?" with "I'm going to be a statistician!"?

Another emotional barrier to the learning of statistics is one related to the difficulty of learning mathematics. The intellectual difficulty of learning mathematics not infrequently creates a phobia currently labeled "math anxiety," a feeling of inadequacy in doing anything mathematical. Such people who dread having to do mathematical work are particularly uncomfortable in the "required" statistics course. We have had many students in our elementary courses who contacted us early in the quarter tell us how nervous they were about passing the course. Many of them tell their history of math anxiety. Some describe their childhood math teachers as particularly stern and unforgiving.

One reason for math anxiety, leading to stat anxiety, is the fact that mathematics is an especially unforgiving *field*. The typical student of arithmetic or elementary algebra doesn't see the beauty and artistry of higher mathematics. Such students see that there is only one correct answer and that no credit is given for approximately correct, or not exactly right but useful. Two plus two equals four, not 4.001. It's either all right or all wrong. In most other subjects, there is a little give: Answers are not so "right" or "wrong."

Those people very facile with numbers can easily intimidate the innumerate. The expert who spews forth a barrage of numerical facts is one up; numerical facts seem more precise and more correct, and the speaker of the facts thus seems more in charge. Robert McNamara, at one time the president of Ford Motor and later the Secretary of Defense under Lyndon Johnson, used to explain the Vietnam war on live television. His command of facts and the confidence with which he spewed forth a barrage of war data gave the impression that he was on top of every detail. He was very impressive, due partly to his command of numbers.

THE THREE LEVELS OF STATISTICAL KNOWLEDGE

Level One

The first level of knowledge is that of familiarity with some of the statistical formulas. These formulas, like the formulas for the t-test or the χ^2 test, are gadgets. They are analogous to the stethoscopes used by physicians, wrenches used by auto mechanics, or desk calculators used by accountants. The formulas of statistics are as important to the field of statistics as the stethoscope, wrench, and desk calculator are for the aforementioned occupations. It's appropriate and natural when learning a new field to play with and get used to its gadgets. It's also a good way to see if you have a chance of liking the field and could be happy working in it. If you can't use a stethoscope, don't consider becoming a physician. If you can't use a wrench, cancel any plans to be an auto mechanic.

The formulas of statistics are qualitatively different from stethoscopes and wrenches. To use a stethoscope requires the ability to find the key spots on the human body, a good ear for subtle sounds, and willingness to tolerate the discomfort of the little black knobs that stick in your ears. Using a wrench requires a good sense of selecting the appropriate size and type of wrench, sufficient arm and shoulder strength to loosen tight nuts, and a light enough touch to avoid ruining a nut. Using statistical formulas requires good facility with algebra at the "college algebra" level, real skill and comfort with mathematical calculations, and a low rate of mistakes in arithmetic.

The skill in using the gadgets of a profession is obviously necessary for its practice. A candidate for the profession is well advised to test the waters by trying to attain at least minimal skill with its gadgets. After attaining minimal skill, the candidate can go on to learn to use them in context.

Level Two

The second level of knowledge of statistics is that of knowing how and when to use what gadgets for standard problems. Just as every physician must know what to do for a patient who is in every way healthy but has just cracked a rib, or every auto mechanic must know how to install a new gas tank, the Level Two statistician knows how to analyze the data from a well-designed household survey when the households are selected using a random number table and every household has an adult respondent at home when the interviewer arrives. The Level Two statistician also knows how to work with a cooperative laboratory researcher who wants to design an experiment to allocate different doses of a chemotherapy to mice with cancer.

Level Three

The Level Three statistician is one who is perfectly familiar with the formulas and can handle difficult, messy problems. A Level Three statistician can assess the possibility of making sense out of large data sets with many missing observations and chaotic methods of collection.

SETTING YOUR OWN GOAL LEVEL

The skill essential to learning statistical formulas is that of working with algebra. In other words, you need to be fairly good with abstract symbols and arithmetic. Being solid in the four basic operations of arithmetic (addition, subtraction, multiplication, division) and taking square roots is absolutely essential for mastery of statistical formulas. Also, being good at looking up numbers in tables is required. There are inexpensive pocket calculators that will do the calculations of statistical formulas, but anyone shaky at arithmetic is in deep statistical water even with calculators. Without the ability to do the calculations by hand, there is great danger of pushing the wrong buttons or pushing the right buttons in the wrong order and not realizing that something is wrong. Another skill essential to the use of statistical formulas is a sense of magnitude, to know when the calculations result in a number that is way too large or too small, or negative when it should be positive. People who confuse debits with credits in their checkbooks have a very hard time with statistical formulas. If you are shaky at arithmetic, don't expect success with even Level One knowledge of statistics.

If you are good at arithmetic and algebra, you should be ready to learn statistical formulas. You would then be ready to follow the instructions of formulas (which you may think of as recipes) and thus correctly perform the calculations.

Learning Level Two requires an additional talent, that of understanding analogies. The ability to see that choosing households for a survey is mathematically equivalent to pulling names out of a hat is essential. Level Two also requires a good memory in order to keep in mind the many situations and where to find the formulas to fit them.

Level Three knowledge of statistics requires, in addition to mastery of the first two levels, the ability and nerve to deal with problems for which there is no standard solution. Working at Level Three requires either finding new solutions or using an old technique that works imperfectly but will do the job. It requires cleverness and adaptability.

TIME REQUIRED FOR LEARNING STATISTICS

For someone good at arithmetic and the symbol manipulations used in algebra, a few of the most popular formulas can be learned in a one-quarter or one-semester elementary course. In a course lasting one academic year, you can cover quite a few of the well-used formulas; you can also start to learn some of the concepts of statistics, although that is a much harder task than learning the formulas.

Learning Level Two takes 2 years of full-time study. This is assuming that you have had a year of calculus (with at least a B grade) and are an advanced under-graduate or graduate student. Level Two is essentially a master's degree in statistics and attaining it is a major effort. Level Three is acquired only after a good 4 years of full-time effort in the field of statistics. It can be obtained by a master's degree plus 2 years full-time experience. Some statisticians attain Level Three by the time they graduate with a Ph.D. in statistics. To attain a very advanced Level Three requires natural statistical talent and many years of experience.

DEALING WITH FORMULAS

People who tend to grimace and flinch at the thought of statistics tend to be most repelled by statistical formulas. We have seen a number of them gingerly open a new statistics text, whose cover beckons the reader with come-ons such as "statistics made easy" or "statistics for the layman." The Introduction assures the reader that he or she need have no background in anything whatsoever and should simply relax and read on. The scarred veteran of previous statistics books doesn't believe it, however, and flips through the book to check for the presence of "formulas." Upon finding them, like bones in a fish, the reader snaps the book shut, defeated once again.

Statistical formulas are frightening to anyone suffering from math anxiety. They remind one of the worst days of the old algebra classes. Some even have Greek letters in them. This book is intended to be light on formulas, but we want to walk the reader through one of the most common ones, just to show how statisticians think about formulas.

There are a couple of well-worn formulas that are used frequently even by non-statisticians. They are the formulas for the two-sample t-test and the chi-square, the latter represented symbolically by the Greek letter chi with a two in the upper right corner: χ^2. By purely arbitrary choice, we'll introduce the former, the two-sample t-test.

A statistics text will typically present the two-sample t-test as follows:

To test $\mathcal{H}_0 : \mu_1 = \mu_2$ vs. $\mathcal{H}_A : \mu_1 \neq \mu_2$ at the α-level, form

$$t = \frac{\overline{x}_1 - \overline{x}_2}{s_p \sqrt{\dfrac{1}{n_1} + \dfrac{1}{n_2}}}$$

where

$$s_p^2 = \frac{\sum_{j=1}^{n_i}(x_{1j} - \overline{x}_1)^2 + \sum_{i=1}^{n_2}(x_{2i} - \overline{x}_2)^2}{n_1 + n_2 - 2}$$

and

$$\overline{x}_i = \frac{\sum_{j=1}^{n_i} x_{ij}}{n_i}$$

for $i = 1,2$. Reject $\mathcal{H}_0$ if and only if

$$|t| > t_{n_1+n_2-2, 1-(\alpha/2)}$$

How are you doing? Had enough? That formula, one of the two most used, is full of frightening symbols and jargon. It contains the Greek letters α, μ and Σ. It has subscripts, indices, and an absolute value symbol.

A statistician can dive right in, plugging in numbers for the x's and looking up the results in something called a t-table. Unless n_1 and n_2 are very small, the statistician may not even need to look in a table, but may know from experience whether t is too large or too small. To the uninitiated, however, the formula is overwhelming. Students in statistics courses need days, sometimes weeks, to get used to it and working with it. Many keep forgetting what μ_1, μ_2, and α are. It seems very artificial, without intuitive appeal. It's intuitively natural to subtract one sample mean from another, i.e., $\overline{x}_1 - \overline{x}_2$, but the whole denominator with s_p and a square root seems weird; it doesn't make any intuitive sense. Students have a hard time keeping standard deviation separate from standard error, forgetting to take square roots, putting the n's in the wrong place. When they should be getting a plus three, they're getting a minus three, and have no instinct that the minus three is wrong. We don't blame them; it is difficult and confusing.

Using statistics formulas is like cooking from recipes or putting together lawn furniture from a set of cut-away drawings. Just as the recipe is a kind of shorthand that could be written in prose form, and the drawings for the swing set could be eliminated by substituting a few paragraphs of English, a statistical formula *could* be expressed in words. A statistician reading the formula does just that, translating the formula into English, seeing it as a set of directions as to what to do with two sets of numbers. For example, we read

$$\overline{x}_i = \frac{\sum_{j=1}^{n_i} x_{ij}}{n_i} \qquad \text{for} \quad i = 1, 2$$

as "add up all the numbers in each of the two groups and divide those totals by the numbers of them in the two groups." The statistician interprets

$$\text{To test } \mathcal{H}_0 : \mu_1 = \mu_2 \text{ vs. } \mathcal{H}_A : \mu_1 \neq \mu_2$$

as "the reason for doing all this arithmetic is to test whether the true averages of the two populations from which the two samples are drawn are exactly equal to each other." It is no surprise, of course, that statisticians write the statistics books in their own language, using formulas instead of English prose because they themselves are so comfortable reading formulas. Formulas, although in principle are similar to recipes, are more abstract. A recipe that says, "Break two eggs into a cup of milk" is referring to objects and actions more common than μ_1 and μ_2, the *conceptual* averages of two populations of numbers. Eggs, cups, and milk can be felt and tasted; μ_1 and μ_2 are abstract concepts. Most people are so used to the small counting numbers, i.e., $1, 2, 3, \ldots$, that they forget how abstract numbers themselves are. Even the number "one" is a figment of the imagination; it cannot be touched or tasted. It has no volume or weight and is not on display in some museum. It is a property that an apple, a truck, and an ocean have in common, namely, their oneness.

In summary, formulas are fine for those very comfortable with algebra and good at following a long list of complicated instructions. If you're weak in either of those areas, don't expect statistical formulas to be easy.

CHAPTER 1

Proportions, Rates, and Ratios

Health decisions are frequently based on proportions, ratios, or rates. In this first chapter we will see how these concepts appeal to common sense and learn their meaning and uses.

1.1. PROPORTIONS

Many outcomes can be classified as belonging to one of two possible categories: Presence and Absence, Non-White and White, Male and Female, Improved and Not-Improved. Of course, one of these two categories is usually identified as of primary interest; for example, Presence in the Presence and Absence classification, Non-White in the White and Non-White classification. We can, in general, re-label the two outcome categories as Positive (+) and Negative (−). An outcome is positive if the primary category is observed and negative if the other category is observed.

It is obvious that in the summary to characterize observations made on a group of individuals, the number x of positive outcomes is not sufficient; the group size n, or total number of observations, should also be recorded. The number x tells us very little and only becomes meaningful after adjusting for the size n of the group; in other words, the two figures x and n are often combined into a *statistic,* called *proportion*:

$$p = \frac{x}{n}$$

(The term *statistic* means a summarized figure from observed data.) Clearly, $0 \leq p \leq 1$. This proportion p is sometimes expressed as a percentage and is calculated as follows:

$$\% = \frac{x}{n}(100)\%$$

Example 1.1

A study, published by the Urban Coalition of Minneapolis and the University of Minnesota Adolescent Health Program, surveyed 12,915 students in grades 7 through 12 in Minneapolis and St. Paul public schools. The report said minority students, about one-third of the group, were much less likely to have *had* a recent routine physical checkup. Among Asian students, 25.4% said they had not seen a doctor or a dentist in the last 2 years, followed by 17.7% of American Indians, 16.1% of blacks, and 10% of Hispanics. Among whites, it was 6.5%.

Proportion is a number used to describe a group of individuals according to a dichotomous characteristic under investigation. The following are a few illustrations of its use in the health sciences.

1.1.1. Comparative Studies

Comparative studies are intended to show possible differences between two or more groups. For example, the same survey of Example 1.1 provided the following figures concerning boys in the surveyed group who use tobacco at least weekly. Among Asians, it was 9.7%, followed by 11.6% of blacks, 20.6% of Hispanics, 25.4% of whites, and 38.3% of American Indians.

Data for comparative studies may come from different sources, with the two fundamental designs being *retrospective* and *prospective*. Retrospective studies gather past data from selected cases and controls to determine differences, if any, in the exposure to a suspected risk factor. They are commonly referred to as *case–control studies*. Cases of a specific disease are ascertained as they arise from population-based registers or lists of hospital admissions, and controls are sampled either as disease-free individuals from the population at risk or as hospitalized patients having a diagnosis other than the one under study. The advantages of a retrospective study are that it is economical and is possible to obtain answers to research questions relatively quickly because the cases are already available. Major limitations are due to inaccuracies in the exposure histories and uncertainty about the appropriateness of the control sample; these problems sometimes hinder retrospective studies and make them less preferred than prospective studies. The following is an example of a retrospective study concerning occupational health.

Example 1.2

A case–control study was undertaken to identify reasons for the exceptionally high rate of lung cancer among male residents of coastal Georgia. Cases were identified from these sources:

1. Diagnoses since 1970 at the single large hospital in Brunswick
2. Diagnoses during 1975–76 at three major hospitals in Savannah
3. Death certificates for the period 1970–74 in the area

Controls were selected from admissions to the four hospitals and from death certificates in the same period for diagnoses other than lung cancer, bladder cancer, or chronic lung cancer. Data are tabulated separately for smokers and non-smokers as follows:

Smoking	Shipbuilding	Cases	Controls
No	Yes	11	35
	No	50	203
Yes	Yes	84	45
	No	313	270

The exposure under investigation, "Shipbuilding," refers to employment in shipyards during World War II.

In an examination of the smokers in the above data set, the numbers of people employed in shipyards, 84 and 45, tell us little because the sizes of the two groups, cases and controls, are different. Adjusting these absolute numbers for the size of the respective group, we have

(i) For the controls,

$$\text{Proportion of exposure} = \frac{45}{315}$$
$$= .143 \text{ or } 14.3\%;$$

(ii) For the cases,

$$\text{Proportion of exposure} = \frac{84}{397}$$
$$= .212 \text{ or } 21.2\%.$$

The results reveal different exposure histories: The proportion among cases was higher than that among controls.

Similar examination of the data for non-smokers shows that, by taking into consideration the numbers of cases and of controls, we have the following figures for employment:

(i) For the controls,

$$\text{Proportion of exposure} = \frac{35}{238}$$
$$= .147 \text{ or } 14.7\%;$$

(ii) For the cases,

$$\text{Proportion of exposure} = \frac{11}{61}$$
$$= .180 \text{ or } 18.0\%.$$

The results also reveal different exposure histories: The proportion among cases was higher than that among controls. The term *exposure* is used here to emphasize that employment in shipyards is a suspected "risk" factor.

The above analyses also show that the difference between proportions of exposure among smokers, that is,

$$21.2 - 14.3 = 6.9\%$$

is different from the difference between proportions of exposure among non-smokers, which is

$$18.0 - 14.7 = 3.3\%$$

In other words, the possible effects of employment in shipyards (as a suspected risk factor) are different for smokers and non-smokers. This difference of differences, if confirmed, is called a *three-term interaction* or an *effect modification*, where smoking alters the effect of employment in shipyards as a risk for lung cancer.

Another example is provided in the following example concerning glaucomatous blindness.

Example 1.3

Persons registered blind from glaucoma

	Population	Cases	Cases per 100,000
White	32,930,233	2,832	8.6
Non-White	3,933,333	3,227	82.0

For these disease registry data, direct calculation of a proportion results in a very tiny fraction, that is, the number of cases of the disease per person at risk. For convenience, this is multiplied by 100,000, and hence the result expresses the number of cases per 100,000 individuals. This data set also provides an example of the use of proportions as disease prevalence, which is defined as

$$\text{Prevalence} = \frac{\text{Number of diseased individuals at the time of investigation}}{\text{Total number of individuals examined}}$$

For blindness from glaucoma, calculations in Example 1.3 reveal a striking difference between the races: The blindness prevalence among non-whites was over eight times that among whites. The number "100,000" was selected arbitrarily; any power of 10 would be suitable so as to obtain a result between 1 and 100 (it is easier to state the result "72 cases per 100,000" than saying that the prevalence was .00072).

1.1.2. Screening Tests

Other uses of proportions can be found in the application of screening tests or diagnostic procedures. Following these procedures, clinical observation or laboratory techniques, individuals are classified as healthy or as falling into one of a number of disease categories. Such tests are important in medicine and epidemiologic studies and form the basis of early interventions. Almost all such tests are imperfect, in the sense that healthy individuals will occasionally be classified wrongly as being ill, while some individuals who are really ill may fail to be detected. Suppose that each individual in a large population can be classified as truly positive or negative for a particular disease; this true diagnosis may be based on more refined methods than are used in the test; or it may be based on evidence that emerges after passage of time, for instance, at autopsy. For each class of individuals, diseased and healthy, the test is applied and results are depicted in Figure 1.1.

The two proportions fundamental to evaluating diagnostic procedures are *sensitivity* and *specificity*. The sensitivity is the proportion of diseased individuals detected as positive by the test

$$\text{Sensitivity} = \frac{\text{Number of diseased individuals who screen positive}}{\text{Total number of diseased individuals}}$$

whereas the specificity is the proportion of healthy individuals detected as negative by the test

$$\text{Specificity} = \frac{\text{Number of healthy individuals who screen negative}}{\text{Total number of healthy individuals}}$$

Clearly, it is desirable that a test or screening procedure should be highly sensitive and highly specific.

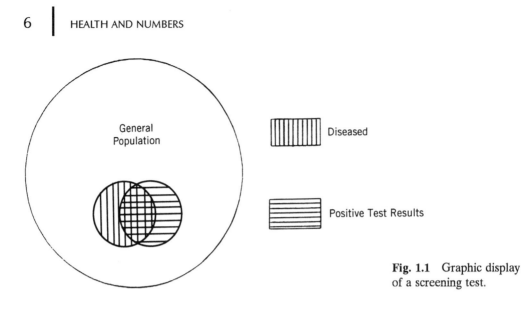

Fig. 1.1 Graphic display of a screening test.

Example 1.4

A cytologic test was undertaken to screen women for cervical cancer. Consider a group of 24,103 women consisting of 379 women whose cervices are abnormal (to an extent sufficient to justify concern with respect to possible cancer) and 23,724 women whose cervices are acceptably healthy. A test was applied and results are tabulated as follows:

	Test		
True	**−**	**+**	**Totals**
−	23,362	362	23,724
+	225	154	379

The calculations

$$\text{Sensitivity} = \frac{154}{379}$$

$$= .406 \text{ or } 40.6\%$$

and

$$\text{Specificity} = \frac{23,362}{23,724}$$

$$= .985 \text{ or } 98.5\%$$

show that the above test is highly specific (98.5%) but not very sensitive (40.6%); there were more than half (59.4%) false negatives. The implications of the use of this test are

(i) If a woman without cervical cancer is tested, the result would almost surely be negative, *but*

(ii) If a woman with cervical cancer is tested, the chance is that the disease would go undetected because 59.4% of these cases would lead to false negatives.

Finally, it is important to note that, throughout this section, proportions have been defined so that both the numerator and the denominator are counts or frequencies, and the numerator corresponds to a subgroup of the larger group involved in the denominator resulting in a number between 0 and 1 (or between 0% and 100%). A straightforward generalization would be in cases with more than two outcome categories, for each category we can define a proportion, and these category-specific proportions add up to 1 (or 100%).

Example 1.5

An examination of the 668 children reported living in crack/cocaine households showed that 70% were black, followed by 18% white, 8% American Indian, and 4% Other or Unknown.

1.1.3. Displaying Proportions

Perhaps the most effective and most convenient way of presenting data, especially discrete data, is through the use of graphs. Graphs convey the information, the general patterns in a set of data, at a single glance. Therefore, graphs are easier to read than tables; the most informative graphs are simple and self-explanatory. Of course, in order to achieve that objective, graphs should be carefully constructed. Like tables, they should be clearly labeled, and units of measurement and/or magnitude of quantities should be included. Remember that graphs must tell their own story; they should be complete in themselves and require little or no additional explanation.

Bar charts

Bar charts are a very popular type of graph used to display several proportions for quick comparison. In a bar chart, the various groups are represented along the horizontal axis; they may be arranged alphabetically, or by the size of their proportions, or on some other rational basis. A vertical bar is drawn above each group such that the height of the bar is the proportion associated with that group. The bars should be of equal width and should be separated from one another so as not to imply continuity.

Pie charts

Pie charts are another popular type of graph. A pie chart consists of a circle; the circle is divided into wedges that correspond to the magnitude of the proportions for various categories. A pie chart shows the differences between the sizes of various categories or subgroups as a decomposition of the total. It is suitable, for example, for use in presenting a budget where we can easily see the difference between expenditures on health care and defense in the United States. Another example of the pie chart's use is for presenting the proportions of deaths due to different causes.

In summary, a bar chart is a suitable graphic device when we have several groups, each associated with a different proportion; whereas a pie chart is more suitable when we have one group that is divided into several categories. The proportions of various categories in a pie chart should add up to 100%. Like bar charts, the categories in a pie chart are usually arranged by the size of the proportions. They may also be arranged alphabetically or on some other rational basis.

Line graphs

A line graph is similar to a bar chart, but the horizontal axis represents time. Different "groups" are consecutive years so that a line graph is suitable to illustrate how certain proportions change over time. In a line graph, the proportion associated with each year is represented by a point at the appropriate height; the points are then connected by straight lines. In addition to their use with proportions, line graphs can also be used to describe changes in the number of occurrences and with continuous measurements.

Example 1.6

1. Kids without a recent physical checkup
   (Refer to data of Example 1.1)

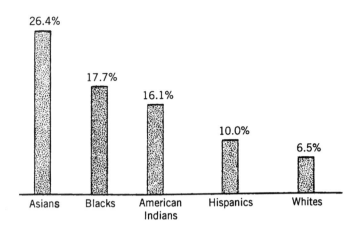

2. <u>Who are the crack kids?</u>
(Refer to data of Example 1.5)

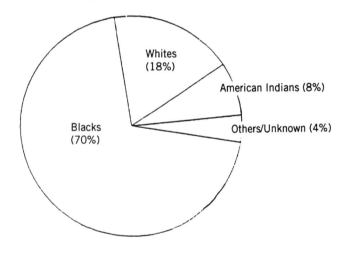

3. <u>Causes of death for Minnesota residents</u>
    The following table provides the number of deaths due to different causes among Minnesota residents for the year 1975.

Cause of death	No. of death
Heart disease	12,378
Cancer	6,448
Cerebrovascular disease	3,958
Accidents	1,814
Others	8,088
Total	32,686

After calculating the proportion of deaths due to each cause, for example,

$$\text{Deaths due to cancer} = \frac{6,448}{32,686}$$

$$= .197 \text{ or } 19.7\%$$

we can present the results as in the following pie chart:

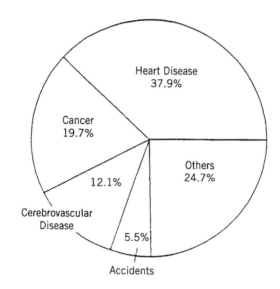

4. Female death rates

Between the years 1984 and 1987, the crude death rates for females in the United States are as follows:

Year	Crude death rate per 100,000 population
1984	792.7
1985	806.6
1986	809.3
1987	813.1

(Figures are proportions × 100,000; see next section for more explanations about these "rates.")

The change in crude death rate for U.S. females can be represented by the following line graph:

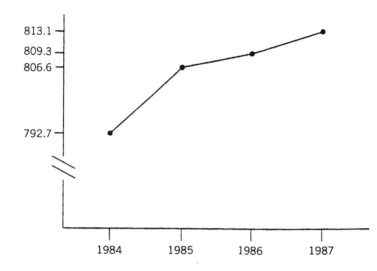

5. Malaria rates in the United States

The following line graph displays the trend in the reported rates of malaria that occurred in the United States between 1940 and 1989 (proportion $\times 100,000$ as above).

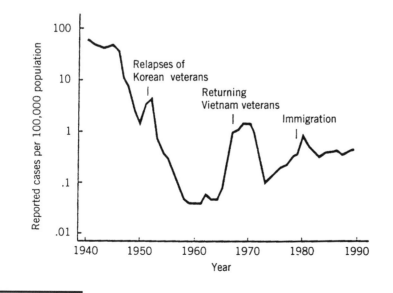

1.2. RATES

In contrast to the static nature of proportions, rates are aimed at measuring the occurrences of events during or after a certain time period.

1.2.1. Changes

Familiar examples of rates include their use to describe changes after a certain period of time.

$$\text{Change rate } (\%) = \frac{\text{New value} - \text{Old value}}{\text{Old value}} \times 100\%$$

In general, change rates could exceed 100%. *They are not proportions* (a proportion is a number between 0 and 1, or 0% and 100%).

Example 1.7

A total of 35,238 new AIDS cases was reported in 1989 by the Centers for Disease Control (CDC) compared to 32,196 reported during 1988. The 9% increase is the smallest since the spread of AIDS began in the early 1980s. For example, new AIDS cases were up 34% in 1988 and 60% in 1987.

In 1989, 547 cases of AIDS transmissions from mothers to newborns were reported, up 17% from 1988. While females made up just 3,971 of the 35,238 new cases reported in 1989, that was an increase of 11% over 1988.

In the above example,

(i) The change rate for new AIDS cases was calculated as

$$\frac{35,238 - 32,196}{32,196} \times 100\% = 9.4\%$$

(ii) For the new AIDS cases transmitted from mothers to newborns, we have

$$17\% = \frac{547 - (1988 \text{ cases})}{(1988 \text{ cases})} \times 100\%$$

leading to

$$1988 \text{ cases} = \frac{547}{1.17}$$
$$= 468$$

Similarly, the number of new AIDS cases for the year 1987 is calculated as follows:

$$34\% = \frac{32{,}196 - (1987 \text{ total})}{(1987 \text{ total})} \times 100\%$$

or

$$1987 \text{ total} = \frac{32{,}196}{1.34}$$
$$= 24{,}027$$

(iii) Among the 1989 new AIDS cases, the proportion of females is

$$\frac{3{,}971}{35{,}238} = .113 \text{ or } 11.3\%$$

and the proportion of males is

$$\frac{35{,}238 - 3{,}971}{35{,}238} = .887 \text{ or } 88.7\%$$

The proportions of females and males add up to 1.0 or 100%.

1.2.2. Measures of Morbidity and Mortality

The field of vital statistics makes some special applications of rates, three kinds of which are commonly used: crude, specific, and adjusted (or standardized). Unlike change rates, these measures are proportions. Crude rates are computed for an entire large group or population; they disregard factors such as age, sex, and race. Specific rates consider these differences among subgroups or categories of diseases. Adjusted or standardized rates are used to make valid summary comparisons between two or more groups possessing different age distributions.

The annual Crude Death Rate is defined as the number of deaths in a calendar year, divided by the population on July 1 of that year; the quotient is often multiplied by 1,000, resulting in a single-digit or double-digit figure. For example, the 1980 population of California was 23,000,000 (as estimated by July 1) and there were 190,237 deaths during 1980, leading to

$$\text{Crude death rate} = \frac{190{,}247}{23{,}000{,}000} \times 1{,}000$$
$$= 8.3 \text{ deaths per 1,000 persons per year}$$

The age-specific and cause-specific death rates are similarly defined.

As for morbidity, the disease prevalence is a proportion used to describe the population at a certain point in time, whereas incidence is a rate used in connection with new cases:

$$\text{Incidence rate} = \frac{\begin{array}{c}\text{Number of individuals who developed the disease}\\\text{over a defined period of time (a year, say)}\end{array}}{\begin{array}{c}\text{Number of individuals initially without the disease}\\\text{who were followed for the defined period of time}\end{array}}$$

For example, the 35,238 new AIDS cases in Example 1.7 and the national population without AIDS at the start of 1989 could be combined according to the above formula to yield an incidence of AIDS for the year.

Another interesting use of rates is in connection with cohort studies. Cohort studies are epidemiologic designs in which one enrolls a group of persons and follows them over certain periods of time; examples include occupational mortality studies. The cohort study design focuses on a particular exposure rather than a particular disease, as in case–control studies. Advantages of a longitudinal approach include the opportunity for more accurate measurement of exposure history and a careful examination of the time relationships between exposure and any disease under investigation. Each member of the cohort belongs to one of three types of termination:

(i) Subjects still alive on the analysis date
(ii) Subjects who died on a known date within the study period
(iii) Subjects who are lost to follow-up after a certain date (these cases are a potential source of bias; effort should be expended on reducing the number of subjects in this category)

The contribution of each member is the length of follow-up time from enrollment to his or her termination. The quotient defined as the observed number of deaths for the cohort, divided by the total follow-up times (in person-years, say) is the rate to characterize the mortality experience of the cohort:

$$\text{Follow-up death rate} = \frac{\text{Number of deaths}}{\text{Total person-years}}$$

Rates may be calculated for total deaths and for separate causes of interest, and they are usually multiplied by an appropriate power of 10, say 1,000, to result in a single-digit or double-digit figure, for example, deaths per 1,000 months of follow-up.

Example 1.8

In an effort to provide a complete analysis of the survival of patients with end-stage renal disease (ESRD), data were collected for a sample that included 929 patients who initiated hemodialysis for the first time at the Regional Disease Program between 1 January 1976 and 30 June 1982; all patients were followed until 31 December 1982. Of these 929 patients, 257 are diabetics; among the 672 nondiabetics, 386 are classified as low risk (without co-morbidities such as arteriosclerotic heart disease, peripheral vascular disease, chronic obstructive pulmonary, and cancer). Results from these two subgroups were as follows:

Group	Age	Deaths/1,000 treatment months
Low risk		
	1–45	2.75
	46–60	6.93
	61+	13.08
Diabetics		
	1–45	10.29
	46–60	12.52
	61+	22.16

For example, for the low-risk patients over 60 years of age, there were 38 deaths during 2,906 treatment months or

$$\frac{38}{2,906} \times 1,000 = 13.08 \text{ deaths per 1,000 treatment months}$$

This is given for illustration only; other data on number of deaths and total treatment months are not given here.

1.2.3. Standardization of Rates

Crude rates, as measures of morbidity or mortality, can be used for population description and may be suitable for investigations of their variations over time; however, the comparisons of crude rates are often invalid because the populations may be different with respect to an important characteristic such as age, sex, or race (called *confounders*). To overcome this difficulty, an adjusted (or standardized) rate is used in the comparison; the adjustment removes the difference in composition with respect to a confounder.

Example 1.9

The following table provides mortality data for Alaska and Florida for the year 1977.

	Alaska			Florida		
Age Group	No. of Deaths	Persons	Deaths per 100,000	No. of Deaths	Persons	Deaths per 100,000
0–4	162	40,000	405.0	2,049	546,000	375.3
5–19	107	128,000	83.6	1,195	1,982,000	60.3
20–44	449	172,000	261.0	5,097	2,676,000	190.5
45–64	451	58,000	777.6	19,904	1,807,000	1,101.5
65+	444	9,000	4,933.3	63,505	1,444,000	4,397.9
Totals	1,615	407,000	396.8	91,760	8,455,000	1,085.3

The above example shows that the 1977 crude death rate per 100,000 population for Alaska was 396.8 and for Florida was 1,085.7. However, a closer examination shows that

(i) Alaska had higher age-specific death rates for four of the five age groups, the only exception being 45–64 years.
(ii) Alaska had a higher percentage of its population in the younger age groups.

The findings make it essential to adjust the death rates of the two states in order to make a valid comparison. A simple way to achieve this, called the *direct method*, is to apply, to a common standard population, age-specific rates observed from the two populations under investigation. For this purpose, the population of the United States as of the last decennial census is frequently used. The procedure consists of the following steps:

1. The standard population is listed by the same age groups.
2. Expected number of deaths in the standard population is computed for each age group of each of the two populations being compared. For example, for age group 0–4, the U.S. population for 1970 was 84,416 (per 1 million); therefore

(i) Alaska rate = 405.0 per 100,000
The expected number of deaths is

$$\frac{(84,416)(405.0)}{100,000} = 341.9$$

$$\cong 342$$

TABLE 1.1 Age-Adjusted Death Rates for Alaska and Florida Using the 1970 U.S. as Standard

Age group	1970 U.S. standard million	Alaska Age-specific rate	Expected deaths	Florida Age-specific rate	Expected deaths
0–4	84,416	405.0	342	375.3	317
5–19	294,353	83.6	246	60.3	177
20–44	316,744	261.0	827	190.5	603
45–64	205,745	777.6	1,600	1,101.5	2,266
65+	98,742	4,933.3	4,871	4,397.9	4,343
Totals	1,000,000		7,886		7,706

(ii) Florida rate = 375.3 per 100,000

The expected number of deaths is

$$\frac{(84,416)(375.3)}{100,000} = 316.8$$

$$\cong 317$$

3. Obtain total number of deaths.
4. Age-adjusted death rate is

$$\text{Adjusted rate} = \frac{\text{Total number of expected deaths}}{\text{Total standard population}} \times 100,000$$

Detailed calculations are given in Table 1.1.

The age-adjusted death rate per 100,000 population for Alaska is 788.6 and for Florida is 770.6. These age-adjusted rates are much closer than as shown by the crude rates. It is important to keep in mind that any population could be chosen as "standard," and, because of this, an adjusted rate is artificial; it does not reflect data from an actual population. The numerical values of the adjusted rates depend in large part on the choice of the standard population. They have real meaning only as relative comparisons.

The advantage of using the U.S. population as the standard is that we can adjust the death rates of many states and compare them with each other. Any population could be selected and used as "standard." In the above example, it does not mean that there were only 1 million people in the U.S. in the year 1970; it only presents the age distribution of 1 million U.S. residents for that year. If all we want to do is to compare Florida versus Alaska, we could choose either one of the states as standard and adjust the death rate of the other; this practice would save half of the labor. For example, if we choose Alaska as the standard population then the adjusted death rate for the state of Florida is calculated as shown in Table 1.2.

TABLE 1.2   Age-Adjusted Death Rates for Florida Using Alaska as Standard

Age group	Alaska population (used as standard)	Florida Rate/100,000	Florida Expected number of deaths
0–4	40,000	375.3	150
5–19	128,000	60.3	77
20–44	172,000	190.5	328
45–64	58,000	1,101.5	639
65+	9,000	4,397.9	396
Totals	407,000		1,590

The new adjusted rate

$$\frac{(1,590)(100,000)}{407,000} = 390.7 \text{ per } 100,000$$

is not the same as that obtained using the 1970 U.S. population as standard (it was 770.6), but it also shows that after age-adjustment the death rate in Florida (390.7 per 100,000) is somewhat lower than in Alaska (396.8 per 100,000; there is no need for adjustment here because we used Alaska's population as the standard population).

1.3. RATIOS

In many cases, such as disease prevalence and disease incidence, proportions and rates are defined very similarly, and the two terms *proportions* and *rates* may even be used interchangeably. Ratio is a completely different concept; it is a computation of the form

$$\text{Ratio} = \frac{a}{b}$$

where *a* and *b* are similar quantities measured from different groups or under different circumstances. An example is the male-to-female ratio of smoking rates; such a ratio may exceed 1.0.

1.3.1. Relative Risk

One of the most often used ratios in epidemiologic studies is the relative risk, a concept for the comparison of two groups or populations with respect to a certain unwanted event (disease or death). The traditional method of expressing it in

**TABLE 1.3   Relative Risks
of Diabetes**

Age group	Relative risk
1–45	3.74
46–60	1.81
61+	1.69

prospective studies is simply the ratio of the incidence rates:

$$\text{Relative risk} = \frac{\text{Disease incidence in group 1}}{\text{Disease incidence in group 2}}$$

However, ratios of disease prevalences as well as follow-up death rates are also applicable. Usually, group 2 is under standard conditions—such as non-exposure to a certain risk factor—against which group 1 (exposed) is measured. For example, if group 1 consists of smokers and group 2 of non-smokers, then we have a relative risk due to smoking. Using the data of Example 1.8, we can obtain the relative risks due to diabetes shown in Table 1.3. All three numbers are greater than 1 and form a decreasing trend with increasing age.

1.3.2. Odds and Odds Ratio

The relative risk is an important index in epidemiologic studies because in such studies it is often useful to measure the increased risk (if any) of incurring a particular disease if a certain factor is present. In cohort studies such an index is readily obtained by observing the experience of groups of subjects with and without the factor as shown above. In a case–control study the data do not present an immediate answer to this type of question, and we now consider how to obtain a useful solution.

Suppose that each subject in a large study, at a particular time, is classified as positive or negative according to some risk factor and as having or not having a certain disease under investigation. For any such categorization the population may be enumerated in a 2×2 table, as shown in Table 1.4. The entries A, B, C, and D in the table are sizes of the four combinations of disease presence and factor presence, and N is the total population size. The relative risk (RR) is

$$RR = \frac{A}{A+B} \div \frac{C}{C+D}$$

$$= \frac{A(C+D)}{C(A+B)}$$

TABLE 1.4   Example of a 2 × 2 Table

Factor	Disease +	Disease −	Total
+	A	B	$A + B$
−	C	D	$C + D$
Total	$A + C$	$B + D$	$N = A + B + C + D$

In many situations, the number of subjects classified as disease positive is small compared with the number classified as disease negative, that is,

$$C + D \cong D$$

$$A + B \cong B$$

and, therefore, the relative risk can be approximated as follows:

$$\text{RR} \cong \frac{AD}{BC}$$

$$= \frac{A/B}{C/D}$$

$$= \frac{A/C}{B/D}$$

(the slash denotes division, and $\cong$ means "almost equal to"). The resulting ratio, AD/BC, is an approximate relative risk, but it is often referred to as an *odds ratio* because

(i) A/B and C/D can be thought of as odds in favor of having disease from groups with or without the factor.
(ii) A/C and B/D can be thought of as odds in favor of exposure to the factors from groups with or without the disease. The two odds can be easily estimated using case–control data, by using sample frequencies.

For the many diseases that are rare, the terms *relative risk* and *odds ratio* are used interchangeably. The relative risk is an important epidemiologic index used to measure seriousness, or the magnitude of the harmful effect of suspected risk factors. For example, if we have

$$\text{RR} = 3.0$$

we can say that the exposed individuals have a risk of contracting the disease that is three times the risk of unexposed individuals. A perfect 1.0 indicates no effect and beneficial factors result in relative risk values that are smaller than 1.0. From data obtained by a case–control or retrospective study it is impossible to calculate the relative risk that we want, but if it is reasonable to assume that the disease is rare (prevalence is less than .05, say) then we can calculate the odds ratio as a "stepping stone" and use it as an approximate relative risk (we use the notation $\cong$ for this purpose). In these cases, we interpret the calculated odds ratio just as we would do with the relative risk.

Example 1.10

The role of smoking in the etiology of pancreatitis has been recognized for many years. To provide estimates of the quantitative significance of these factors, a hospital-based study was carried out in eastern Massachusetts and Rhode Island between 1975 and 1979. Ninety-eight patients who had a hospital discharge diagnosis of pancreatitis were included in this unmatched case–control study. The control group consisted of 451 patients admitted for diseases other than those of the pancreas and biliary tract. Risk factor information was obtained from a standardized interview with each subject, conducted by a trained interviewer.

The following are some data for the males:

Use of cigarettes	Cases	Controls
Never	2	56
Ex-smokers	13	80
Current smokers	38	81
	53	217

In Example 1.10, the approximate relative risks or odds ratios are

(i) For ex-smokers,

$$RR_e \cong \frac{13/2}{80/56}$$

$$= \frac{(13)(56)}{(80)(2)}$$

$$= 4.55$$

(The subscript e in RR_e indicates that we are calculating the relative risk (RR) for **ex**-smokers.)

(ii) For current smokers,

$$RR_c \cong \frac{38/2}{81/56}$$

$$= \frac{(38)(56)}{(81)(2)}$$

$$= 13.14$$

In these calculations, the nonsmokers are used as references. These values indicate that the risk of having pancreatitis for current smokers is approximately 13.14 times the same risk for people who never smoke. The effect for ex-smokers is smaller (4.55 times) but it is still very high (as compared to 1.0—the no-effect baseline for relative risks and odds ratios).

1.3.3. Standardized Mortality Ratio

In a cohort study, the follow-up death rates are calculated and used to describe the mortality experience of the cohort under investigation. However, the observed mortality of the cohort is often compared with that expected from the death rates of the national population. The basis of this method is the comparison of the observed number of deaths, d, from the cohort with the mortality that would have been expected if the group had experienced similar death rates to those of the national population of which the cohort is a part. Let e denote the expected number of deaths; then the comparison is based on the following ratio, called the *standardized mortality ratio* (SMR):

$$SMR = \frac{d}{e}$$

The expected number of deaths is calculated using published national life tables, and the calculation can be approximated as follows:

$$e \cong \lambda T$$

where T is the total follow-up time (person-years) from the cohort and λ the annual death rate (per person) from the referenced population. Of course, the annual death rate of the referenced population changes with age. Therefore, what we actually do in research is more complicated, although based on the same idea. First, we subdivide the cohort into many age groups, then calculate the product λT for each age group using the correct age-specific rate for that group, and add up the results.

Example 1.11

Some 7,000 British workers exposed to vinyl chloride monomer were followed for several years to determine whether their mortality experience differed from that of the general population. The following data are for deaths from cancers and are tabulated separately for four groups based on years since entering the industry. This data set shows some interesting features:

(i) For the group with 1–4 years since entering industry, we have a death rate that is substantially less than that of the general population (SMR = .445 or 44.5%). This phenomenon, known as the *healthy worker* effect, is most likely a consequence of a selection factor whereby workers are necessarily in good health at the time of their entry into the work force.

(ii) We see an attenuation of the healthy worker effect with the passage of time, so that the cancer death rates show a slight excess after 15 years (vinyl chloride exposure is known to induce a rare form of liver cancer and to increase rates of brain cancer).

Deaths from cancers	Years since entering the industry				Total
	1–4	5–9	10–14	15+	
Observed	9	15	23	68	115
Expected	20.3	21.3	24.5	60.8	126.8
SMR (%)	44.5	70.6	94.0	111.8	90.7

Taking the ratio of two standardized mortality ratios is another way of expressing relative risk. For example, the relative risk of the 15+ years group is 1.58 times the risk of the risk of the 5–9 years group, since the ratio of the two corresponding mortality ratios is

$$\frac{111.8}{70.6} = 1.58$$

Similarly, the risk of the 15+ years group is 2.51 times the risk of the 1–4 years group because the ratio of the two corresponding mortality ratios is

$$\frac{111.8}{44.5} = 2.51$$

EXERCISES

1. Self-reported injuries among left-handed and right-handed people were compared in a survey of 1,896 college students in British Columbia, Canada. Of the 180 left-handed students, 93 reported at least one injury, and 619 of the 1,716 right-handed students reported at least one injury in the same period. Calculate the proportion of people with at least one injury during the period of observation for each group. Arrange the data into a 2×2 table.

2. A study was conducted to evaluate the hypothesis that tea consumption and premenstrual syndrome are associated. One hundred eighty-eight nursing students and 64 tea factory workers were given questionnaires. The prevalence of premenstrual syndrome was 39% among the nursing students and 77% among the tea factory workers. How many people in each group have premenstrual syndrome?

3. The relationship between prior condom use and tubal pregnancy was assessed in a population-based case–control study at Group Health Cooperative of Puget Sound during 1981–86. The results are

Condom use	Cases	Controls
Never	176	488
Ever	51	186

Compute the proportion of subjects in each group who never used condoms.

4. The following table provides the proportions of currently married women having an unplanned pregnancy; data are tabulated for several different methods of contraception.

Method of contraception	Proportion with pregnancy
None	0.431
Diaphragm	0.149
Condom	0.106
IUD	0.071
Pill	0.037

Display these proportions in a bar chart.

5. The following table summarizes the coronary heart disease (CHD) and lung cancer mortality rates per 1,000 person-years by number of cigarettes smoked per day at baseline for men participating in MRFIT.

		CHD Deaths		Lung cancer deaths	
	Total	N	rate/1,000 yr	N	rate/ 1,000 yr
Never-smokers	1,859	44	2.22	0	0
Ex-smokers	2,813	73	2.44	13	0.43
Smokers					
1–19 cig/day	856	23	2.56	2	0.22
20–39 cig/day	3,747	173	4.45	50	1.29
≥ 40 cig/day	3,591	115	3.08	54	1.45

For each cause of death, display the rates in a bar chart.

6. The following table provides data taken from a study on the association between race and use of medical care by adults experiencing chest pain in the past year.

Response	Black	White
MD seen in past year	35	67
MD seen, not in past year	45	38
MD never seen	78	39
Total	158	144

Display the proportions for each group in a separate pie chart.

7. The following frequency distribution provides the number of cases of pediatric AIDS between 1983 and 1989.

Year	No. of cases
1983	122
1984	250
1985	455
1986	848
1987	1,412
1988	2,811
1989	3,098

Display the trend of numbers of cases using a line graph.

8. A study was conducted to investigate the changes between 1973 and 1985 in women's use of three preventive health services. The data were obtained from the National Health Survey; women were divided into subgroups according to

age and race. The percentages of women receiving a breast examination within the past 2 years are given below.

	Breast exam within past 2 years	
Age and race	1973	1985
Total	65.5	69.6
Black	61.7	74.8
White	65.9	69.0
20–39 years	77.5	77.9
Black	77.0	83.9
White	77.6	77.0
40–59	62.1	66.0
Black	54.8	67.9
White	62.9	65.7
60–79 years	44.3	56.2
Black	39.1	64.5
White	44.7	55.4

Separately for each group, blacks and whites, display the proportions of women receiving a breast examination within the past 2 years in a bar chart. Mark the midpoint of each age group on the horizontal axis and display the same data using a line graph.

9. Consider the following data:

	Tuberculosis		
X-ray	No	Yes	Total
Negative	1,739	8	1,747
Positive	51	22	73
Total	1,790	30	1,820

Calculate the sensitivity and specificity of x-ray as a screening test for tuberculosis.

10. Sera from a T-lymphotropic virus type (HTLV-1) risk group (prostitute women) were tested with two commercial "research" enzyme-linked immunoabsorbent assays (EIA) for HTLV-1 antibodies. These results were compared with a gold standard, and outcomes are shown in the following table.

True	Dupont's EIA		Cellular Product's EIA	
	Positive	Negative	Positive	Negative
Positive	15	1	16	0
Negative	2	164	7	179

Calculate and compare the sensitivity and specificity of the two EIA's.

11. The following table provides the number of deaths for several leading causes among Minnesota residents for the year 1991.

Cause of death	No. of deaths	Rate per 100,000 population
Heart disease	10,382	294.5
Cancer	8,299	?
Cerebrovascular disease	2,830	?
Accidents	1,381	?
Other causes	11,476	?
Total	34,368	?

(a) Calculate the percent of total deaths for deaths from each cause and display the results in a pie chart.

(b) From death rate (per 100,000 population) for heart disease, calculate the population for Minnesota for the year 1991.

(c) From the result of (b), fill in the missing death rates (per 100,000 population) at the question marks in the above table.

12. The survey described in Example 1.1 continues in the Comparative Studies section providing percentages of boys from various ethnic groups who use tobacco at least weekly. Display these proportions in a bar chart similar to the one in Example 1.6 (2).

13. A case–control study was conducted relating to the epidemiology of breast cancer and the possible involvement of dietary fats, along with other vitamins and nutrients. It included 2,024 breast cancer cases who were admitted to Roswell Park Memorial Institute, Erie County, New York, from 1958 to 1965. A control group of 1,463 was chosen from the patients having no neoplasms and no pathology of gastrointestinal or reproductive systems. The primary factors being investigated were vitamins A and E (measured in international units per month). The following are data for 1,500 women over 54 years of age.

Vitamin A (IU/mo)	Cases	Controls
$\leq 150{,}500$	893	392
$> 150{,}500$	132	83
Total	1,025	475

Calculate the odds ratio associated with a decrease in ingestion of foods containing vitamin A.

14. Refer to the data set in Example 1.2.

   (a) Calculate the odds ratio associated with employment in shipyards for non-smokers.

   (b) Calculate the same odds ratio for smokers.

   (c) Compare the results of (a) and (b); a large difference would indicate a "three-term interaction" or "effect modification," where smoking alters the effect of employment in shipyards as a risk for lung cancer.

15. Although cervical cancer is not a major cause of death among American women, it has been suggested that virtually all such deaths are preventable. In an effort to find out who is being screened for the disease, data from the 1973 National Health Interview (a sample of the United States population) were used to examine the relationship between Pap testing and some socioeconomic factors. The following table provides the percentages of women who reported never having had a Pap test (these are from metropolitan areas):

Age	Income	White (%)	Black (%)
25–44	Poor	13.0	14.2
	Nonpoor	5.9	6.3
45–64	Poor	30.2	33.3
	Nonpoor	13.2	23.3
65 and over	Poor	47.4	51.5
	Nonpoor	36.9	47.4

   (a) Calculate the odds ratios associated with race (black vs. white) among

   (i) 25–44 Nonpoor

   (ii) 45–64 Nonpoor

   (iii) 65+ Nonpoor

   Briefly discuss a possible effect modification if any.

   (b) Calculate the odds ratios associated with income (Poor vs. Nonpoor) among

   (i) 25–44 Black

(ii) 45–64 Black

(iii) 65+ Black

Briefly discuss a possible effect modification if any.

(c) Calculate the odds ratios associated with race (Black vs. White) among

(i) 65+ Poor

(ii) 65+ Nonpoor

Briefly discuss a possible effect modification.

16. An important characteristic of glaucoma, an eye disease, is the presence of classic visual field loss. Tonometry is a common form of glaucoma screening, wherein, for example, an eye is classified as positive if it has an intra-ocular pressure of 21 mmHg or higher at a single reading. Given the following data,

Field	Test result		
loss	Positive	Negative	Total
Yes	13	7	20
No	413	4,567	4,980

calculate the sensitivity and specificity of this screening test.

17. From information given in Example 1.7, calculate

(a) The number of new AIDS cases for the years 1987 and 1986

(b) The number of cases of AIDS transmission from mothers to newborns for 1988

18. For the low-risk ESRD patients in Example 1.8 we also had the following follow-up data:

Age (years)	Deaths	Treatment months
21–30	4	1,012
31–40	7	1,387
41–50	20	1,706
51–60	24	2,448
61–70	21	2,060
Over 70	17	846

Compute the follow-up death rate for each age group and the relative risk for group "over 70" versus "51–60."

19. The following are mortality data for the State of Georgia for the year 1977:

Age group	Deaths	Population
0–4	2,483	424,600
5–19	1,818	1,818,000
20–44	3,656	1,126,500
45–64	12,424	870,800
65+	21,405	360,800

Using these and the data in Example 1.9, calculate the crude and age-adjusted death rates for Georgia and compare them to those for Alaska, the U.S. population given in Example 1.9 being used as the standard. Calculate again with the Alaska population serving as the standard population.

20. The study mentioned in Example 1.11 also provided the following data for deaths due to circulatory disease:

	Years since entering the industry				
Deaths	1–4	5–9	10–14	15+	Total
Observed	7	25	38	110	180
Expected	32.5	35.6	44.9	121.3	234.1

Calculate the SMRs for each subgroup and the relative risk for group "15+" versus group "1–4."

21. A long-term follow-up study of diabetes has been conducted among Pima Indian residents of the Gila River Indian Community of Arizona since 1965. Subjects of this study, at least 5 years old and of at least half Pima ancestry, were examined approximately every 2 years; examinations included measurements of height and weight and a number of other factors. The following table relates diabetes incidence rate (new cases/1,000 person-years) to body mass index (a measure of obesity defined as weight/[height]^2).

Body mass index	Incidence rate
< 20	.8
20–25	10.9
25–30	17.3
30–35	32.6
35–40	48.5
≥ 40	72.2

Display these rates by means of a bar chart.

22. In the course of selecting controls for a study to evaluate effect of caffeine-containing coffee on the risk of myocardial infarction among women 30–49 years of age, a study noted appreciable differences in coffee consumption among hospital patients admitted for illnesses not known to be related to coffee use. Among potential controls, the coffee consumption of patients who had been admitted to the hospital by conditions having an acute onset (such as fractures) was compared with that of patients admitted for chronic disorders.

| | Cups of coffee per day | | | |
Admission by	0	1–4	≥ 5	Total
Acute conditions	340	457	183	980
Chronic conditions	2,440	2,527	868	5,835

Find a way to express this consumption difference by means of a graph.

23. In a seroepidemiologic survey of health workers representing a spectrum of exposure to blood and patients with hepatitis B virus (HBV), it was found that infection increased as a function of contact. The following table provides data for hospital workers with uniform socioeconomic status at an urban teaching hospital in Boston, Massachusetts.

Personnel	Exposure	n	HBV positive
Physicians	Frequent	81	17
	Infrequent	89	7
Nurses	Frequent	104	22
	Infrequent	126	11

(a) Calculate the proportion of HBV-positive workers in each subgroup.

(b) Calculate the odds ratios associated with frequent contacts (as compared with infrequent contacts); do this separately for physicians and nurses.

(c) Compare the two ratios obtained in (b); a large difference would indicate a "three-term interaction" or "effect modification," where effects of "frequency" are different for physicians and nurses.

24. The results of the Third National Cancer survey have shown substantial variation in lung cancer incidence rates for white males within Allegheny County,

Pennsylvania, which may be due to different smoking rates. The following table gives the percentages of current smokers by age for two study areas.

Age	Lawrenceville		South Hills	
	n	%	n	%
35–44	71	54.9	135	37.0
45–54	79	53.2	193	28.5
55–64	119	43.7	138	21.7
≥ 65	109	30.3	141	18.4
Total	378	46.8	607	27.1

(a) Display the age distribution for Lawrenceville by means of a pie chart.

(b) Display the age distribution for South Hills by means of a bar chart.

(c) Find a way to show the difference in age distribution by means of a graph.

25. Prematurity, which ranks as the major cause of neonatal morbidity and mortality, has traditionally been defined on the basis of a birth weight under 2,500 g. But this definition encompasses two distinct types of infants: infants who are small because they are born early and infants who are born at or near term but are small because their growth was retarded. *Prematurity* has now been replaced by

(i) *Low birth weight* to describe the second type

(ii) *Preterm* to characterize the first type (babies born before 37 weeks of gestation)

A case–control study of the epidemiology of preterm delivery was undertaken at Yale–New Haven Hospital in Connecticut during 1977. The study population consisted of 175 mothers of singleton preterm infants and 303 mothers of singleton full-term infants. The following tables give the distributions of age and socioeconomic status.

Age	Cases	Controls
14–17	15	16
18–19	22	25
20–24	47	62
25–29	56	122
≥ 30	35	78

Socio-economic level	Cases	Controls
Upper	11	40
Upper middle	14	45
Middle	33	64
Lower middle	59	91
Lower	53	58
Unknown	5	5

Find a way (or ways) to summarize data so as to express the observation that preterm mothers are younger and poorer.

26. Sudden infant death syndrome (SIDS), also known as *sudden unexplained death*, *crib death*, or *cot death*, claims the lives of an alarming number of apparently normal infants every year. In a study at the University of Connecticut School of Medicine, significant associations were found between SIDS and certain demographic characteristics. Some of the summarized data are given below:

	No. of deaths	
	Observed	Expected
Sex		
Male	55	45
Female	35	45
Race		
Black	23	11
White	67	79

(Expected deaths are calculated using Connecticut infant mortality data for 1974–1976.)

(a) Calculate the SMR for each subgroup.

(b) Compare males versus females and blacks versus whites.

27. Adult male residents of 13 counties in western Washington state in whom testicular cancer had been diagnosed during 1977–1983 were interviewed over the telephone regarding their history of genital tract conditions, including vasectomy. For comparison, the same interview was given to a sample of men selected from the population of these counties by dialing telephone numbers at random. The following data are tabulated by religious background.

Religion	Vasectomy	Cases	Controls
Protestant	Yes	24	56
	No	205	239
Catholic	Yes	10	6
	No	32	90
Others	Yes	18	39
	No	56	96

Calculate the odds ratio associated with vasectomy for each religious group. Is there any evidence of an effect modification?

28. In 1979 the U.S. Veterans Administration conducted a health survey of 11,230 veterans. The advantages of this survey are that it includes a large random sample with a high interview response rate and it was done before the recent public controversy surrounding the issue of the health effects of possible exposure to Agent Orange. The following are data relating Vietnam service to eight post-traumatic stress disorder symptoms among the 1,787 veterans who entered the military service between 1965 and 1975.

	Service in Vietnam	
Symptom	Yes	No
Nightmares		
Yes	197	85
No	577	925
Sleep problems		
Yes	173	160
No	599	851
Troubled memories		
Yes	220	105
No	549	906
Depression		
Yes	306	315
No	465	699
Temper control problems		
Yes	176	144
No	595	868
Life goal association		
Yes	231	225
No	539	786
Omit feelings		
Yes	188	191
No	583	821
Confusion		
Yes	163	148
No	607	864

Calculate the odds ratio for each symptom.

29. The roles of menstrual and reproductive factors in the epidemiology of breast cancer have been reassessed using pooled data from three large case–control studies of breast cancer from several Italian regions. The following are summarized data for age at menopause and age at first live birth.

Age	Cases	Controls
At first live birth		
< 22	621	898
22–24	795	909
25–27	791	769
≥ 28	1,043	775
At menopause		
< 45	459	543
45–49	749	803
≥ 50	1,378	1,167

Find a way (or ways) to summarize the data further so as to express the observation that the risk of breast cancer is lower for women with younger ages at first live birth and younger ages at menopause.

30. It has been hypothesized that dietary fiber decreases the risk of colon cancer, while meats and fats are thought to increase this risk. A large study was undertaken to confirm these hypotheses. Fiber and fat consumptions are classified as "low" or "high" and data are tabulated separately for males and females as follows ("low" means below median):

	Males		Females	
Diet	Cases	Controls	Cases	Controls
Low fat, high fiber	27	38	23	39
Low fat, low fiber	64	78	82	81
High fat, high fiber	78	61	83	76
High fat, low fiber	36	28	35	27

For each group (males and females), using "low fat, high fiber" as the baseline, calculate the odds ratio associated with each other dietary combination. Any evidence of an effect modification (interaction between consumption of fat and consumption of fiber)?

2

Summarization of
Continuous Measurements

A class of measurements or a characteristic on which individual observations or measurements are made is called a *variable*; examples include weight, height, and blood pressure. Suppose we have a set of numerical values for a variable:

(i) If each element of this set may lie only at a few isolated points, we have a *discrete* data set. Examples are race, sex, counts of events, or some sort of artificial grading.

(ii) If each element of this set may theoretically lie anywhere on the numerical scale, we have a *continuous* data set. Examples are blood pressure, cholesterol level, or time to a certain event such as death.

The previous chapter deals with the summarization and description of discrete data; in this chapter the emphasis is on continuous measurements.

2.1. TABULAR AND GRAPHICAL METHODS

There are different ways of organizing and presenting data; simple tables and graphs, however, are still very effective methods. They are designed to help the reader obtain an intuitive feeling for the data at a glance.

2.1.1. Frequency Distribution

There is no difficulty if the data set is small, for we can arrange those few numbers and write them, say, in increasing order; the result would be sufficiently clear. For fairly large data sets, a useful device for summarization is the formation of a *frequency table* or *frequency distribution*. This is a table showing the number of observations, called *frequency*, within certain ranges of values of the variable under investigation. For example, taking the variable to be the age at death, we have the following example; the second column of the table provides the frequencies.

Example 2.1

The following table gives the number of deaths by age for the state of Minnesota in 1987.

Age	No. of deaths
Less than 1	564
1–4	86
5–14	127
15–24	490
25–34	667
35–44	806
45–54	1,425
55–64	3,511
65–74	6,932
75–84	10,101
85 and over	9,825
Total	34,524

If a data set is to be grouped to form a frequency distribution, difficulties should be recognized and an efficient strategy is needed for better communication. First, there is no clear-cut rule on the number of intervals or classes. With too many intervals, the data are not summarized enough for a clear visualization of how they are distributed. On the other hand, too few intervals are undesirable because the data are oversummarized and some of the details of the distribution may be lost. In general, between 5 and 15 intervals are acceptable; of course, this also depends on the number of observations.

The widths of the intervals must also be decided. Example 2.1 shows the special case of mortality data, where it is traditional to show infant deaths (deaths of persons who are born live but die before living 1 year). Without such specific reasons, intervals generally should be of the same width. This common width w may be determined by dividing the range R by k, the number of intervals:

$$w = \frac{R}{k}$$

where the range R is the difference between the smallest and the largest numbers in the data set. In addition, a width should be chosen so that it is convenient to use or easy to recognize, such as a multiple of 5 (or 1 if the data set has a narrow range). Similar considerations apply to the choice of the beginning of the first interval; it is a convenient number that is low enough for the first interval to include the smallest observation. Finally, care should be taken in deciding in which interval to place an observation falling on one of the interval boundaries. For example, a consistent rule could be made so as to place such an observation in the interval of which the observation in question is the lower limit.

Example 2.2

The following are weights in pounds of 57 children at a daycare center:

68	63	42	27	30	36	28	32	79	27
22	23	24	25	44	65	43	25	74	51
36	42	28	31	28	25	45	12	57	51
12	32	49	38	42	27	31	50	38	21
16	24	69	47	23	22	43	27	49	28
23	19	46	30	43	49	12			

From the above data set we have

1. The smallest number is 12 and the largest is 79 so that

$$R = 79 - 12$$
$$= 67$$

If five intervals are used we would have

$$w = \frac{67}{5}$$
$$= 13.4$$

and if 15 intervals are used we would have

$$w = \frac{67}{15}$$
$$= 4.5$$

Between these two values, 4.5 and 13.6, there are two convenient numbers: 5 and 10. Since the sample size of 57 is not large, a width of 10 should be an apparent choice because it results in few intervals.

2. Since the smallest number is 12, we may begin our first interval at 10. These considerations, (1) and (2), lead to the following seven intervals:

10–19

20–29

30–39

40–49

50–59

60–69

70–79

3. Determining the frequencies or the number of values or measurements for each interval is merely a matter of examining the values one by one and of placing a tally mark beside the appropriate interval. When we do this we have the following table.

TABLE 2.1   Frequency Distribution of Weights of 57 Children

Weight interval (lb.)	Tally	Frequency (%)	Relative Frequency (%)
10–19	卌	5	8.8
20–29	卌 卌 卌 ‖‖	19	33.3
30–39	卌 卌	10	17.5
40–49	卌 卌 ‖‖	13	22.8
50–59	‖‖‖	4	7.0
60–69	‖‖‖	4	7.0
70–79	‖	2	3.5
Total		57	100.0

4. An optional step in the formulation of a frequency distribution is to present the proportion or *relative frequency*, in addition to frequency, for each interval. These proportions, defined by

$$\text{Relative frequency} = \frac{\text{Frequency}}{\text{Total number of observations}}$$

are shown in the third column of Table 2.1 (as %) and would be very useful if we need to compare two data sets of different sizes.

Example 2.3

A study was conducted to investigate the possible effects of exercise on the menstrual cycle. From the data collected from that study, we obtained the menarchal age (in years) of 56 female swimmers who began their swimming training after they had reached menarche; these served as controls in order to compare with those who began their training prior to menarche.

14.0	16.1	13.4	14.6	13.7	13.2	13.7	14.3
12.9	14.1	15.1	14.8	12.8	14.2	14.1	13.6
14.2	15.8	12.7	15.6	14.1	13.0	12.9	15.1
15.0	13.6	14.2	13.8	12.7	15.3	14.1	13.5
15.3	12.6	13.8	14.4	12.9	14.6	15.0	13.8
13.0	14.1	13.8	14.2	13.6	14.1	14.5	13.1
12.8	14.3	14.2	13.5	14.1	13.6	12.4	15.1

From this data set we have

1. The smallest number is 12.4 and the largest is 16.1 so that

$$R = 16.1 - 12.4$$

$$= 3.7$$

If five intervals are used, we would have

$$w = \frac{3.7}{5}$$

$$= .74$$

and if 15 intervals are used, we would have

$$w = \frac{3.7}{15}$$

$$= .25$$

Between these two values, .25 and .74, .5 seems to be a convenient number to use as the width; .25 is another choice, but it would create many intervals (15) for such a small data set. (Another alternative is to express ages in months and not to deal with decimal numbers.)

2. Since the smallest number is 12.4, we may begin our intervals at 12.0, leading to the following intervals:

> 12.0–12.4
>
> 12.5–12.9
>
> 13.0–13.4
>
> 13.5–13.9
>
> 14.0–14.4
>
> 14.5–14.9
>
> 15.0–15.4
>
> 15.5–15.9
>
> 16.0–16.4

3. Count the number of swimmers whose ages belong to each of the above nine intervals, the frequencies, and obtain the following table—completed with the last column for relative frequencies (expressed as %):

Age (years)	Frequency (%)	Relative frequency (%)
12.0–12.4	1	1.8
12.5–12.9	8	14.3
13.0–13.4	5	8.9
13.5–13.9	12	21.4
14.0–14.4	16	28.6
14.5–14.9	4	7.1
15.0–15.4	7	12.5
15.5–15.9	2	3.6
16.0–16.4	1	1.8
Total	56	100.0

2.1.2. Histogram and the Frequency Polygon

A convenient way of displaying a frequency table is by means of a *histogram* and/or a *frequency polygon*. A histogram is a diagram in which

1. The horizontal scale represents the value of the variable marked at interval boundaries

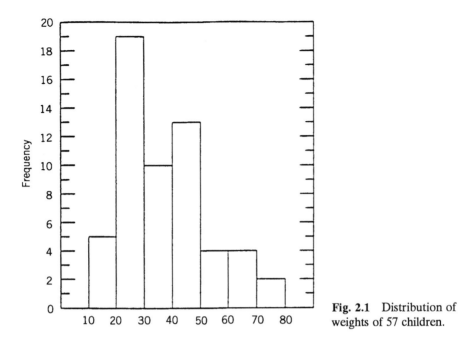

Fig. 2.1 Distribution of weights of 57 children.

2. The vertical scale represents the frequency or relative frequency in each interval (when intervals or classes have unequal widths, this scale represents *densities*; see definition, below)

The histogram presents us with a graphic picture of the distribution of measurements. This picture consists of rectangular bars joining each other, one for each interval as shown in Figure 2.1 for the data set of Example 2.2. If disjoint intervals are used such as in Table 2.1, the horizontal axis is marked with true boundaries. For example, 19.5 serves as the true upper boundary of the first interval and true lower boundary for the second interval. In cases where we need to compare the shapes of the histograms representing different data sets, the height of each rectangular bar should represent the density of the interval, where the interval density is defined by

$$\text{Density} = \frac{\text{Relative frequency}}{\text{Interval width}} \times 100\%$$

If we do this, the relative frequency is represented by the area of the rectangular bar, and the total area under the histogram is 100%. It may be a good practice to always graph densities on the vertical axis with or without having equal class width; when class widths are equal, the shape of the histogram looks similar to the graph with relative frequencies on the vertical axis.

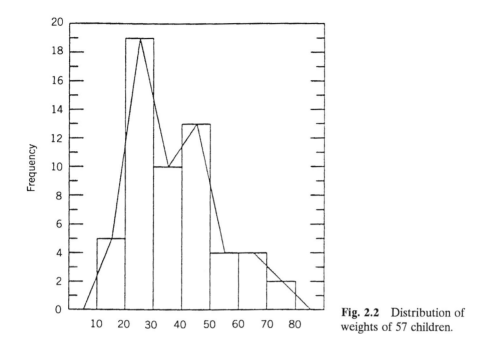

Fig. 2.2 Distribution of weights of 57 children.

To draw a frequency polygon, we first place a dot at the midpoint of the upper base of each rectangular bar. The points are connected with straight lines. At the ends, the points are connected to the midpoints of the previous and succeeding intervals (these are make-up intervals with zero frequency, where widths are the widths of the first and last intervals, respectively). A frequency polygon as thus constructed is another way to portray graphically the distribution of a data set (see Figure 2.2). The frequency polygon can also be shown without the histogram on the same graph.

The frequency table and its graphic relatives the histogram and the frequency polygon have a number of applications, as explained below; the first leads to a research question, and the second leads to a new analysis strategy.

1. When data are homogeneous, the table and graphs usually show a uni-modal pattern with one peak in the middle part. A bimodal pattern might indicate possible influence or effect of certain hidden factors.

Example 2.4

The following table provides data on age and percentage saturation of bile for 31 male patients.

Age	% Saturation	Age	% Saturation	Age	% Saturation
23	40	55	137	48	78
31	86	31	88	27	80
58	111	20	88	32	47
25	86	23	65	62	74
63	106	43	79	36	58
43	66	27	87	29	88
67	123	63	56	27	73
48	90	59	110	65	118
29	112	53	106	42	67
26	52	66	110	60	57
64	88				

Using 10% intervals, the above data set can be represented by a histogram or a frequency polygon as shown in Figure 2.3. This picture shows an apparent bimodal distribution; however, a closer examination shows that among the 9 patients with over 100% saturation, 8 (or 89%) are over 50 years of age. On the other hand, only 4 (or 18%) of 22 patients with less than 100 percent saturation are over 50 years of age. The two peaks in the diagram correspond to the two age groups.

2. Another application concerns the symmetry of the distribution as depicted by the table or its graphs. A symmetric distribution is one in which the distribution has the same shape on both sides of the peak location. If there are more extremely large values, the distribution is then skewed to the right, or *positively* skewed. Examples include family income, antibody level after vaccination, and drug dose to produce a predetermined level of response, among others. It is common that,

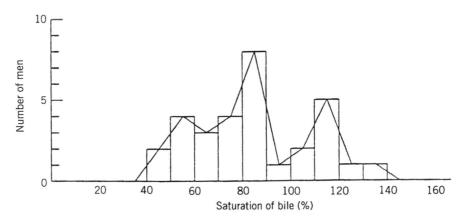

Fig. 2.3 Frequency polygon for percentage saturation of bile in men.

for positively skewed distributions, subsequent statistical analyses should be performed on the log scale, for example, to compute and/or to compare averages of log(dose).

___

Example 2.5

The distribution of family income for the United States in 1983 by race is shown below.

| | Percent of families | |
Income ($)	White	Non-White
0–14,999	13	34
15,000–19,999	24	31
20,000–24,999	26	19
25,000–34,999	28	13
35,000–59,999	9	3
60,000 and over	1	Negligible
Total	100	100

___

The distribution for non-white families is represented in the histogram in Figure 2.4, where the vertical axis represents the density (percent per thousand dollars). It is obvious that the distribution is not symmetric; it is very skewed to the right.

In this histogram, we graph the densities on the vertical axis. For example, for the second income interval ($15,000–19,999), the relative frequency is 31% and the width of the interval is $5,000 (31% per $5,000), leading to the density,

$$\frac{31}{5,000} \times 1,000 = 6.2$$

or 6.2% per $1,000 (we arbitrarily multiply by 1,000—or any power of 10—just to obtain a larger number for easy graphing).

2.1.3. Cumulative Frequency Graph and Percentiles

Cumulative relative frequency, or *cumulative percentage*, gives the percentage of individuals having a measurement less than or equal to the upper boundary of the class interval. Data from Table 2.1 are reproduced and supplemented with a column for cumulative relative frequency in Table 2.2.

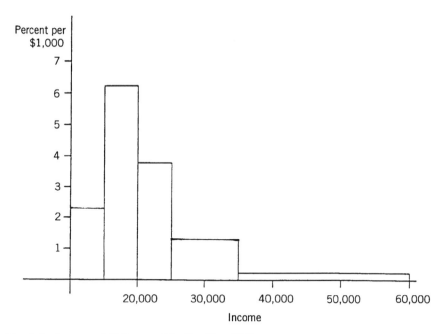

Fig. 2.4 Income of U.S. non-white families, 1983.

TABLE 2.2   Distribution of Weights of 57 Children

Weight interval (lbs)	Frequency	Relative frequency (%)	Cumulative relative frequency (%)
10–19	5	8.8	8.8
20–29	19	33.3	42.1
30–39	10	17.5	59.6
40–49	13	22.8	82.4
50–59	4	7.0	89.4
60–69	4	7.0	96.4
70–79	2	3.5	99.9 ≅ 100.0
Total	57	100.0	

This last column is easy to form; you do it by successively cumulating the relative frequencies of each of the various intervals. In the example in Table 2.2, the cumulative percentage for the first three intervals is

$$8.8 + 33.3 + 17.5 = 59.6$$

and we can say that 59.6% of the children in the data set have a weight of 39.5 lbs or less. Or, as another example, 96.4% of children weigh 69.5 lbs or less, and so on.

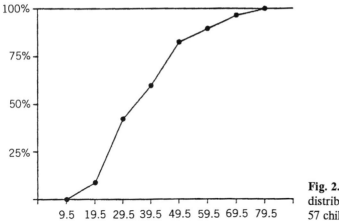

Fig. 2.5 Cumulative distribution of weights of 57 children.

The cumulative relative frequency can be presented graphically as in Figure 2.5. This type of curve is called a *cumulative frequency graph*. To construct a cumulative frequency graph, we place a point with the horizontal axis marked at the upper class boundary and vertical axis marked at the corresponding cumulative frequency. Each point represents the cumulative relative frequency for that interval, and the points are connected with straight lines. At the left end, it is connected to the lower boundary of the first interval. If disjoint intervals, such as

10–19

20–29

etc...

are used, points are placed at the true boundaries.

The cumulative percentages and their graphical representation the cumulative frequency graph have a number of applications.

1. When two cumulative frequency graphs, representing two different data sets, are placed on the same graph, they provide a rapid visual comparison without any need to compare individual intervals. Figure 2.6 gives such a comparison of family incomes using data of Example 2.4.

2. The cumulative frequency graph provides a class of important statistics known as *percentiles* or *percentile scores*. The 90th percentile, for example, is that observation that exceeds 90% of the values in the data set and is exceeded by only 10% of them. Or, as another example, the 80th percentile is that numerical value that exceeds 80% of the values contained in the data set and is exceeded by 20% of them, and so on. The 50th percentile is commonly called the *median*. In our

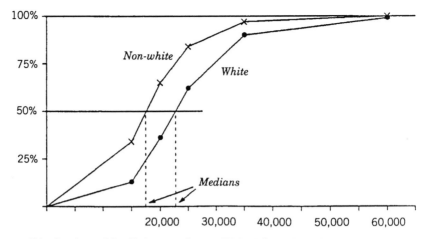

Fig. 2.6 Distributions of family income for the United States in 1983.

example of Figure 2.6, the median family income in 1983 for non-whites was about $17,500 as compared to a median of about $22,000 for white families. To get the median, we start at the 50% point on the vertical axis and go horizontally until meeting the cumulative frequency graph; the projection of this intersection on the horizontal axis is the median. Other percentiles are obtained similarly.

The cumulative frequency graph also provides an important application in the formation of health norms (see Figure 2.7) for the monitoring of physical progress (weight and height) of infants and children. Here, the same percentiles, say 90th, of weight or height of groups of different ages are joined by a curve.

Example 2.6

Figure 2.8 provides results from a study of Hmong refugees in the Minneapolis–St. Paul area where each dot represents the average height of five refugee girls of the same age. The graph shows that even though the refugee girls are small, mostly in the lowest 25%, they grow at the same rate as measured by the American standard. However, the pattern changes by the age of 15 years, when their average height drops to a level below the 5th percentile. In this example, we plot the *average* height of five girls instead of individual heights; because the Hmongs are small, individual heights are likely to be out of the chart. This concept of *average* or *mean* will be further explained in Section 2.2.

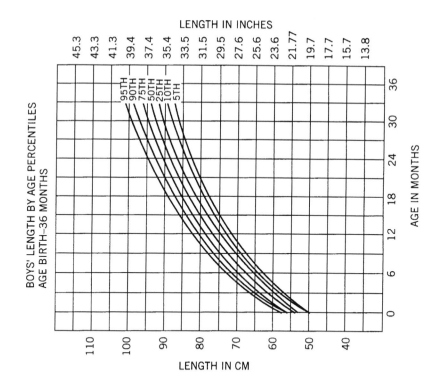

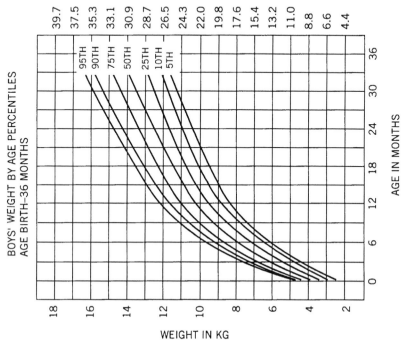

Fig. 2.7 Height (right) and weight (left) curves.

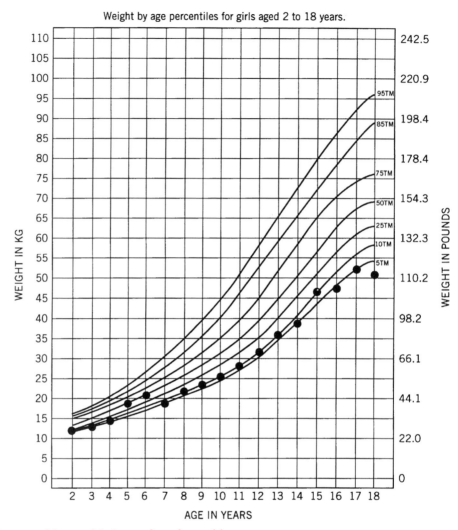

Fig. 2.8 Mean weight by age for refugee girls.

2.2. NUMERICAL METHODS

Although tables and graphs serve useful purposes, there are many situations that require other types of data summarization. What we need in many applications is the ability to summarize data by means of just a few numerical measures, particularly before inferences or generalizations are drawn from the data. Measures for describing the location (or typical value) of a set of measurements and their variation or dispersion are used for these purposes.

First, let us suppose we have n measurements in a data set; for example, here is a data set

$$\{8, 2, 3, 5\}$$

with $n = 4$. We usually denote these numbers as x_i; thus we have for the above example: $x_1 = 8$, $x_2 = 2$, $x_3 = 3$, and $x_4 = 5$. If we add all the x_i in the above data set, we obtain 18 as the sum. This addition process is recorded as

$$\sum x = 18$$

where the Greek letter Σ is the summation sign.

With the summation notation, we are now able to define a number of important summarized measures starting with the arithmetic average or *mean*.

2.2.1. Mean

Given a data set of size n,

$$\{x_1, x_2, \ldots, x_n\}$$

the mean of the x's will be denoted by $\bar{x}$ ("x-bar") and is computed by summing all the x's and dividing the sum by n. Symbolically,

$$\bar{x} = \frac{\sum x}{n}$$

It is important to know that Σ ("sigma") stands for an operation (that of obtaining the sum of the quantities that follow) rather than a quantity itself. For example, considering the data set

$$\{8, 5, 4, 12, 15, 5, 7\}$$

we have

$$n = 7$$

$$\sum x = 56$$

leading to

$$\bar{x} = \frac{56}{7}$$

$$= 8$$

Occasionally, data—especially second-hand data—are presented in the grouped form of a frequency table. In these cases, the mean $\bar{x}$ can be approximated using

TABLE 2.3 Calculation of the Mean $\bar{x}$ Using Frequency Table 2.2

Weight interval	Frequency (f)	Interval midpoint (m)	fm
10–19	5	14.5	72.5
20–29	19	24.5	465.5
30–39	10	34.5	345.0
40–49	13	44.5	578.5
50–59	4	54.5	218.0
60–69	4	64.5	258.0
70–79	2	74.5	149.0
Totals	57		2,086.5

the following formula ($\cong$ means "approximately equal to"):

$$\bar{x} \cong \frac{\sum(fm)}{n}$$

where f denotes the frequency, that is, the number of observations in an interval; m the interval midpoint; and the summation is across the intervals. The midpoint for an interval is obtained by calculating the average of the interval lower true boundary and the upper true boundary. For example, if the first three intervals are

10–19

20–29

30–39

then the midpoint for the first interval is

$$\frac{9.5 + 19.5}{2} = 14.5$$

and for the second interval is

$$\frac{19.5 + 29.5}{2} = 24.5$$

This process is illustrated in Table 2.3.

$$\bar{x} \cong \frac{2,086.5}{57}$$

$$= 36.6 \text{ lbs}$$

(If individual weights were used, we would have $\bar{x} = 36.7$ lbs.)

Of course, the mean $\bar{x}$ obtained from this technique with a frequency table is different from the $\bar{x}$ using individual or raw data. However, the process saves some computational labor and the difference between the results, $\bar{x}$'s, is very small if the data set is large and the interval width is small.

As earlier indicated, a characteristic of some interest is the symmetry or lack of symmetry of a distribution, and it was recommended that for very positively skewed distributions analyses are commonly done on the log scale. After obtaining a mean on the log scale, you should take the antilog to return to the original scale of measurement; the result is called the *geometric mean* of the x's. The effect of this process is to minimize the influences of extreme observations (very large numbers in the data set). For example, considering the data set

$$\{8, 5, 4, 12, 15, 7, 28\}$$

with one unusually large measurement, we have the following table with natural logs presented in the second column:

	x	$\ln x$
	8	2.08
	5	1.61
	4	1.39
	12	2.48
	15	2.71
	7	1.95
	28	3.33
Totals:	79	15.55

The mean is

$$\bar{x} = \frac{79}{7}$$

$$= 11.3$$

while on the log scale, we have

$$\frac{\sum \ln x}{n} = \frac{15.55}{7}$$

$$= 2.22$$

leading to a geometric mean of 9.22, which is less affected by the large measurements. Geometric mean is used extensively in microbiologic and serologic research in which distributions are often positively skewed.

2.2.2. Other Measures of Location

Another useful measure of location is the *median*. If the observations in the data set are arranged in increasing or decreasing order, the median is the middle observation that divides the set into equal halves. If the number of observations n is odd, there will be a unique median, the $\frac{1}{2}(n+1)$th number from either end in the ordered sequence. If n is even, there is strictly no middle observation, but the median is defined by convention as the average of the two middle observations, the $\frac{1}{2}n$th and $\frac{1}{2}(n+1)$th from either end. In Section 2.1, we showed a quicker way to get an approximate value for the median using the cumulative frequency graph (see Figure 2.6).

The above two small data sets have different means, 8 and 11, but the same median, 8. Therefore, the advantage of the median as a measure of location is that it is less affected by extreme observations. However, the median has some disadvantages in comparison with the mean:

1. It takes no account of the precise magnitude of most of the observations and is therefore less efficient than the mean because it wastes information.
2. If two groups of observations are pooled, the median of the combined group cannot be expressed in terms of the medians of the two component groups. However, the mean can be so expressed. If component groups are of sizes n_1 and n_2 and have means $\bar{x}_1$ and $\bar{x}_2$, respectively, the mean of the combined group is

$$\bar{x} = \frac{n_1 \bar{x}_1 + n_2 \bar{x}_2}{n_1 + n_2}$$

3. In large data sets, the median requires more work to calculate than the mean and is not much use in the elaborate statistical techniques (it is still useful as a descriptive measure for skewed distributions).

A third measure of location, the *mode*, was briefly introduced in Section 2.1.2. It is a value at which the frequency polygon reaches a peak. The mode is not widely used in analytical statistics, other than as a descriptive measure, mainly because of the ambiguity in its definition, as the fluctuations of small frequencies are apt to produce spurious modes. Because of these reasons, the rest of this book is focused on only one measure of location, the mean.

2.2.3. Measures of Dispersion

When the mean $\bar{x}$ of a set of measurements has been obtained it is usually a matter of considerable interest to measure the degree of variation or dispersion around this mean. Are the x's all rather close to $\bar{x}$, or are some of them dispersed widely in each direction? This question is important for purely descriptive reasons, but it is also important because the measurement of dispersion or variation plays a central part in the methods of statistical inference, which will be described in subsequent chapters of this book.

An obvious candidate for the measurement of dispersion is the range R, defined as the difference between the largest value and the smallest value, which was introduced in Section 2.1.2. However, there are a few difficulties about the use of the range. The first is that the value of the range is determined by only two of the original observations. Second, the interpretation of the range depends, in a complicated way, on the number of observations, which is an undesirable feature.

An alternative approach is to make use of the *deviations* from the mean, $x - \bar{x}$; it is obvious that the greater the variation in the data set, the larger the magnitude of these deviations tends to be. From these deviations, the *variance* s^2 is computed by squaring each deviation, adding them, and dividing their sum by one less than n:

$$s^2 = \frac{\sum(x - \bar{x})^2}{n - 1}$$

The use of the divisor $(n - 1)$ instead of n is clearly not very important when n is large. It is more important for small values of n, and its justification will be briefly explained later in this section. It should be noted that

1. It would be no use to take the mean of deviations because

$$\sum(x - \bar{x}) = 0$$

2. Taking the mean of the absolute values, e.g.,

$$\frac{\sum|x - \bar{x}|}{n}$$

is a possibility. However, this measure has the drawback of being difficult to handle mathematically, and we will not consider it further in this book.

The variance s^2 (s-squared) is measured in the square of the units in which the x's are measured. For example, if x is the time in seconds, the variance is measured in square seconds, (seconds)^2. It is convenient, therefore, to have a measure of variation expressed in the same units as the x's, and this can be easily done by taking the square root of the variance. This quantity is known as the *standard deviation*, and its formula is

$$s = \sqrt{\frac{\sum(x - \bar{x})^2}{n - 1}}$$

Consider again the data set

$$\{8, 5, 4, 12, 15, 5, 7\}$$

Calculation of the variance s^2 and standard deviation s is illustrated in Table 2.4.

In general, this calculation process is likely to cause some trouble. If the mean is not a "round" number, say $\bar{x} = 10/3$, it will need to be rounded off, and errors arise

TABLE 2.4   Calculation of Variance and Standard Deviation

x	$x - \bar{x}$	$(x - \bar{x})^2$
8	0	0
5	−3	9
4	−4	16
12	4	16
15	7	49
5	−3	9
7	−1	1
$\sum x = 56$		$\sum(x - \bar{x})^2 = 100$
$n = 7$		$s^2 = 100/6 = 16.67$
$\bar{x} = 8$		$s = \sqrt{16.67} = 4.08$

TABLE 2.5   Use of a Short-Cut Formula for Variance

x	x^2
8	64
5	25
4	16
12	144
15	225
5	25
7	49
56	548

in the subtraction of this figure from each x. This difficulty can be easily overcome with use of the following short-cut formula for the variance:

$$s^2 = \frac{\sum x^2 - \frac{(\sum x)^2}{n}}{n - 1}$$

The previous example is re-worked in Table 2.5.

$$s^2 = \frac{(548) - (56)^2/7}{6}$$

$$= 16.67$$

When data are presented in the grouped form of a frequency table, the variance is calculated using the following modified short-cut formula ($\cong$ means "approxi-

TABLE 2.6 Calculation of the Variance Using Frequency Table 2.2

Weight interval	f	m	m^2	fm	fm^2
10–19	5	14.5	210.25	72.5	1,051.25
20–29	19	24.5	600.25	465.5	11,404.75
30–39	10	34.5	1,190.25	345.0	11,902.50
40–49	13	44.5	1,980.25	578.5	25,743.25
50–59	4	54.5	2,970.25	218.0	11,881.00
60–69	4	64.5	4,160.25	258.0	16,641.00
70–79	2	74.5	5,550.25	149.0	11,100.50
Totals	57			2,086.5	89,724.25

$$s^2 \cong \frac{89{,}724.25 - (2{,}086.5)^2/57}{56}$$

$$= 238.35$$

$$s \cong 15.4 \text{ lbs}$$

(If individual weights were used, we would have $s = 15.9$ lbs.)

mately equal to"):

$$s^2 \cong \frac{\sum fm^2 - \dfrac{(\sum fm)^2}{n}}{n - 1}$$

where f denotes an interval frequency, m the interval midpoint calculated as in the previous section, and the summation is across the intervals. This approximation is illustrated in Table 2.6.

It is often not clear to beginners why we use $(n - 1)$ instead of n as the denominator for s^2. This number $(n - 1)$ is called the *degrees of freedom* representing the *amount of information* contained in the sample. The real explanation for $(n - 1)$ is hard to present at the level of this text; however, it may be seen this way. What we are trying to do with s is to provide a measure of variability, a measure of the "average gap" or "average distance" between numbers in the sample—and there are $(n - 1)$ "gaps" between n numbers.

Finally, it is occasionally useful to describe the variation by expressing the standard deviation as a proportion or percentage of the mean. The resulting measure

$$CV = \frac{s}{\bar{x}} \times 100\%$$

is called the *coefficient of variation*. It is a dimensionless quantity because the standard deviation is expressed in the same units as the mean and could be used to compare the difference in variation between two types of measurements. However, its use is rather limited, and we will not present it at this level.

2.3. LIFE-TABLE METHODS

2.3.1. Survival Data

In some studies, the important number is the time to an event, such as death; it is called the *survival time*. The term *survival time* is conventional even though the primary event could be nonfatal, such as a relapse of a disease or the appearance of the first disease symptom. What is special about the analysis of survival data is that at the time of data analysis there are still subjects without the event. These subjects contain only partial information, for example, having survived 10 years or more, and are called *censored* cases. The censored cases are important featues of survival data; with their presence all the methods of summarization we learned—such as $\bar{x}$ and s—become irrelevant. It is impossible to calculate $\bar{x}$, because the true survival time of a censored case is unknown.

Example 2.7

The remission times of 42 patients with acute leukemia were reported from a clinical trial undertaken to assess the ability of the drug 6-mercaptopurine (6-MP) to maintain remission. Each patient was randomized to receive either 6-MP or placebo. The study was terminated after 1 year; patients have different follow-up times because they were enrolled sequentially at different times. Times in weeks were

---

6-MP group
   6, 6, 6, 7, 10, 13, 16, 22, 23, 6+, 9+, 10+, 11+, 17+, 19+, 20+, 25+, 32+, 32+, 34+, 35+

Placebo group
   1, 1, 2, 2, 3, 4, 4, 5, 5, 8, 8, 8, 8, 11, 11, 12, 12, 15, 17, 22, 23

---

in which a $t+$ denotes a censored observation, i.e., the case was censored after t weeks without a relapse. For example, "10+" is a case enrolled 10 weeks before study termination and still in remission at termination.

In the analysis of survival data, an important statistic that is of considerable interest is the "t-year survival rate" (in the context of the above sample, it is the

t-week survival rate). The t-year survival rate is defined as

$$\frac{\text{Number of individuals who survive longer than } t \text{ years}}{\text{Total number of individuals in the data set}}$$

The 5-year survival rate is commonly used in cancer research as a measure of a treatment effectiveness. In clinical trials, the differences between survival rates are indications of a treatment effect. Without censoring, the calculation of survival rates is straightforward; for example, in the above placebo group,

$$10\text{-week survival rate} = \frac{8}{21} \times 100\%$$
$$= 38.1\%$$

because 8 of 21 cases survived 10 weeks or more. In the following sections we will discuss methods appropriate for censored data. The main objective is to take into account censored cases that contain only partial information.

2.3.2. Kaplan-Meier Curve

We first introduce the *product-limit* (PL) method of estimating the survival rates; this is also called the *Kaplan-Meier* method. According to this method, survival rates are calculated by constructing a table such as Table 2.5, with five columns following the outlines below:

1. Column 1 contains all the survival times in the data set, both censored and un-censored, in order from smallest to largest. If a censored observation has the same value as an uncensored one, the latter appears first.
2. The second column consists of ranks from 1 to n.
3. The third column is essentially the same as the second but only for uncensored observations; values are labeled as r.
4. Compute the term

$$\frac{n - r}{n - r + 1}$$

   for every r of column 3.
5. To obtain the survival rate at time t, multiply all values in column 4 up to and including t. If some uncensored observations are ties, the smallest rate is the answer. From Table 2.7, we have, for example:

6-week survival rate is 85.72%

22-week survival rate is 53.78%

The results of Table 2.7 show that, for example, 62.75% of patients are still in re-mission after 16 weeks. Of course, if we did not use censored cases, we would underestimate this 16-week survival rate (from relapse).

TABLE 2.7 Estimation of Survival Rates for the 6-MP Group in Example 2.7

(1)	(2)	(3)	(4)	(5)
t	r		$\dfrac{n-r}{n-r+1}$	Percent still in remission (t-week survival rate)
6	1	1	.9524	.9524
6	2	2	.9500	.9048
6	3	3	.9474	$\rightarrow$ .8572
6+	4	—	—	—
7	5	5	.9412	.8068
9+	6	—	—	—
10	7	7	.9333	.7530
10+	8	—	—	—
11+	9	—	—	—
13	10	10	.9167	.6903
16	11	11	.9091	.6275
17+	12	—	—	—
19+	13	—	—	—
20+	14	—	—	—
22	15	15	.8571	$\rightarrow$ .5378
23	16	16	.8333	.4482
25+	17	—	—	—
32+	18	—	—	—
32+	19	—	—	—
34+	20	—	—	—
35+	21	—	—	—

A convenient graphical way of displaying survival rates is by means of a *survival curve* or *Kaplan-Meier curve*. A survival curve is a graph in which

1. The horizontal scale represents the times marked at uncensored observations
2. The vertical scale represents the survival rates

so that each uncensored observation is represented by a dot and these dots are joined so as to form a step curve. For the data of Example 2.5, survival rates for the two groups are displayed in the same graph as shown in Figure 2.9.

2.3.3. Actuarial Method

The Kaplan-Meier method is applicable for any survival data set; however, for a large data set—say 100 or more patients—it may be much more convenient to group the times into intervals. The process is similar to the formation of a frequency table, and the method is referred to as the *actuarial method*. A concrete example is adapted (Table 2.8) to illustrate this method with eight columns following the out-

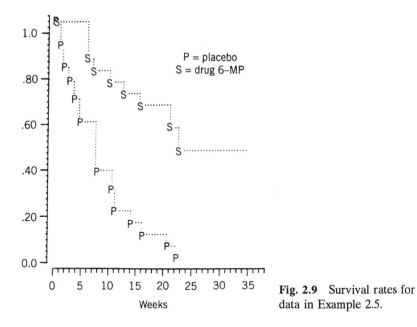

Fig. 2.9 Survival rates for data in Example 2.5.

TABLE 2.8   Survival of Patients With a Particular Form of Malignant Disease

(1)	(2)	(3)	(4)	(5)	(6)	(7)	(8)
	Last reported		Living at the	No. at	Interval death	Interval survival	Survival rate at
Time (years) (t to $[t + 1]$)	Died (d)	Censored (c)	start (n)	risk (n')	rate (q)	rate (p)	t years (%)
0–1	90	0	374	374.0	.2406	.7594	100.0
1–2	76	0	284	284.0	.2676	.7324	75.9
2–3	51	0	208	208.0	.2452	.7548	55.6
3–4	25	12	157	151.0	.1656	.8344	42.0
4–5	20	5	120	117.5	.1702	.8298	35.0
5–6	7	9	95	90.5	.0773	.9227	29.1
6–7	4	9	79	74.5	.0537	.9463	26.8
7–8	1	3	66	64.5	.0155	.9845	25.4
8–9	3	5	62	59.5	.0504	.9496	25.0
9–10	2	5	54	51.5	.0388	.9612	23.7
10+	21	26	47	—	—	—	22.8

lines below:

1. The choice of the time intervals will depend on the nature of the data and on the size of the data set; guidelines given in Section 2.1.1 (Frequency Distribution) are equally applicable here.

2, 3. The patients are classified according to the time interval during which their condition was last reported. If the report was a death, the patient is counted in column (2) and if censored the patient is counted in column (3); time for each patient is calculated from his or her enrollment in the study.

4. The number of patients living at the start of the intervals is obtained by cumulating columns (2) and (3) from the bottom. For example, the number alive at 10 years is $21 + 26 = 47$ and the number alive at 9 years is $47 + 2 + 5 + 54$, and so on.

5. The number of patients at risk during interval t to $(t + 1)$ is

$$n' = n - c/2$$

The purpose of this formula is to provide a denominator for the next column. The adjustment from n to n' is needed because the c censored subjects are necessarily at risk for only part of the interval. The basic assumption here is that, on the average, these patients are at risk for half of the interval.

6. The interval death rate is

$$q = d/n'$$

For example, during the first interval

$$q = \frac{90}{374.0}$$

$$= .2406$$

7. The interval survival rate

$$p = 1 - q$$

8. To obtain the survival rate at t years, multiply all values in column (7) up to and including that for interval t to $(t + 1)$. For example,

$$\text{1-year survival rate} = 75.9\%$$

$$\text{5-year survival rate} = 29.1\%$$

Survival rate calculated by the actuarial method can also be displayed by the survival curve similar to that for rates obtained by the Kaplan-Meier method. The only difference here is that the dots in the diagram are connected by straight lines as shown in Figure 2.10.

Finally, it is noted that this approximation—the actuarial method—is still useful even if the user has access to computer packages for constructing the Kaplan-Meier curve because it provides an estimate for the "hazard" or "risk" function, which is not possible with the Kaplan-Meier method and which is useful and needed for certain more advanced analyses.

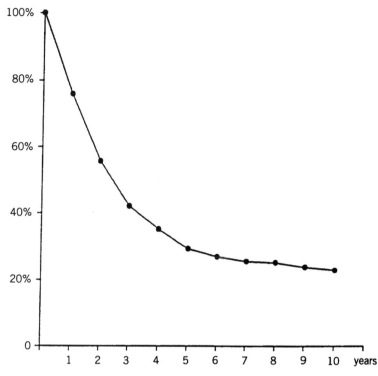

Fig. 2.10 Survival curve for cancer patients in Table 2.8.

EXERCISES

1. The following table gives the values of serum cholesterol levels for 1,067 U.S. men, aged 25 to 34 years.

Cholesterol level (mg/100 ml)	No. of men
80–119	13
120–159	150
160–199	442
200–239	299
240–279	115
280–319	34
320–399	14
Total	1,067

(a) Plot the histogram, the frequency polygon, and the cumulative frequency graph.

(b) Find the median.

2. The following table provides the relative frequencies of blood lead concentrations for two groups of workers in Canada, one examined in 1979 and the other in 1987.

Blood lead (µg/dl)	Relative frequency (%)	
	1979	1987
0–20	11.5	37.8
20–29	12.1	14.7
30–39	13.9	13.1
40–49	15.4	15.3
50–59	16.5	10.5
60–69	12.8	6.8
70–79	8.4	1.4
80+	9.4	0.4

(a) Plot the histogram and frequency polygon for each year.

(b) Plot the two cumulative frequency graphs in one figure.

(c) Find and compare the medians.

3. A study on the effects of exercise on the menstrual cycle provides the following ages (years) of menarche (the beginning of menstruation) for 96 female swimmers who began training prior to menarche:

15.0	17.1	14.6	15.2	14.9	14.4	14.7	15.3
13.6	15.1	16.2	15.9	13.8	15.0	15.4	14.9
14.2	16.5	13.2	16.8	15.3	14.7	13.9	16.1
15.4	14.6	15.2	14.8	13.7	16.3	15.1	14.5
16.4	13.6	14.8	15.5	13.9	15.9	16.0	14.6
14.0	15.1	14.8	15.0	14.8	15.3	15.7	14.3
13.9	15.6	15.4	14.6	15.2	14.8	13.7	16.3
15.1	14.5	13.6	15.1	16.2	15.9	13.8	15.0
15.4	14.9	16.2	15.9	13.8	15.0	15.4	14.9
14.2	16.5	13.4	16.5	14.8	15.1	14.9	13.7
16.2	15.8	15.4	14.7	14.3	15.2	14.6	13.7
14.9	15.8	15.1	14.6	13.8	16.0	15.0	14.6

(a) Form a frequency distribution including relative frequencies and cumulative relative frequencies.

(b) Plot the frequency polygon and the cumulative frequency graph.

(c) Find the median and 95th percentile.

4. The following are the menarchal ages (in years) of 56 female swimmers who began training after they had reached menarche.

14.0	16.1	13.4	14.6	13.7	13.2	13.7	14.3
12.9	14.1	15.1	14.8	12.8	14.2	14.1	13.6
14.2	15.8	12.7	15.6	14.1	13.0	12.9	15.1
15.0	13.6	14.2	13.8	12.7	15.3	14.1	13.5
15.3	12.6	13.8	14.4	12.9	14.6	15.0	13.8
13.0	14.1	13.8	14.2	13.6	14.1	14.5	13.1
12.8	14.3	14.2	13.5	14.1	13.6	12.4	15.1

(a) Form a frequency distribution using the same age intervals as in Exercise 1. (These intervals may be different from those in Example 2.3.)

(b) Display in the same graph two cumulative frequency graphs: training before menarche and training after menarche. Any conclusions?

(c) Find the median and 95th percentile and compare them to the results of the previous exercise.

5. The following table shows the daily fat intake (grams) of a group of 150 adult males.

22	62	77	84	91	102	117	129	137	141
42	56	78	73	96	105	117	125	135	143
37	69	82	93	93	100	114	124	135	142
30	77	81	94	97	102	119	125	138	142
46	89	88	99	95	100	116	121	131	152
63	85	81	94	93	106	114	127	133	155
51	80	88	98	97	106	119	122	134	151
52	70	76	95	107	105	117	128	144	150
68	79	82	96	109	108	117	120	147	153
67	75	76	92	105	104	117	129	148	164
62	85	77	96	103	105	116	132	146	168
53	72	72	91	102	101	128	136	143	164
65	73	83	92	103	118	127	132	140	167
68	75	89	95	107	111	128	139	148	168
68	79	82	96	109	108	117	130	147	153

(a) Form a frequency distribution including relative frequencies and cumulative relative frequencies.

(b) Plot the frequency polygon and investigate the symmetry of the distribution.

(c) Plot the cumulative frequency graph and find the 25th and 75th percentiles. Also calculate the *mid-range* = 75th percentile − 25th percentile; this is another good descriptive measure of variation.

6. Using the income data in Example 2.5,

(a) Plot and compare the two cumulative frequency graphs: whites and non-whites.

(b) Compute and compare the medians.

7. Refer to the percentage saturation of bile for the 31 male patients in Example 2.4.

(a) Compute the mean, the variance (using short-cut formula), and the standard deviation.

(b) The frequency polygon of Figure 2.3 is based on the following grouping (arbitrary choices):

Interval (%)	Frequency
40–49	2
50–59	4
60–69	3
70–79	4
80–89	8
90–99	1
100–109	2
110–119	5
120–129	1
130–139	1
140–149	0

Plot the cumulative frequency graph and obtain the median. How does the answer compare to the exact median (the 16th largest saturation percentage)?

8. The same study referenced in Example 2.4 also provided data (percentage saturation of bile) for 29 women. These percentages were

65	58	52	91	84	107
86	98	35	128	116	84
76	146	55	75	73	120
89	80	127	82	87	123
142	66	77	69	76	

(a) Form a frequency distribution using the same intervals as in Exercise 5.

(b) Plot in the same graph and compare the two frequency polygons and cumulative frequency graphs: men and women.

(c) Compute the mean, the variance, and the standard deviation.

(d) Compute and compare the two coefficients of variation: men and women.

9. The following frequency distribution was obtained for the pre-operational percentage hemoglobin values of a group of subjects from a village where there has been a malaria eradication program (MEP):

Hemoglobin (%)	30–39	40–49	50–59	60–69	70–79	80–89	90–99
Frequency	2	7	14	10	8	2	2

The results in another was obtained after an MEP and are given below:

43	63	63	75	95	75	80	48	62	71	76	90
51	61	74	103	93	82	74	65	63	53	64	67
80	77	60	69	73	76	91	55	65	69	84	78
50	68	72	89	75	57	66	79	85	70	59	71
87	67	72	52	35	67	99	81	97	74	61	62

(a) Form a frequency distribution using the same intervals as in the first table.

(b) Plot in the same graph and compare the two cumulative frequency graphs: before and after the malaria eradication program.

10. In a study of water pollution, a sample of mussels was taken and lead concentration (milligrams per gram dry weight) was measured from each one. The following data were obtained:

$$\{113.0, 140.5, 163.3, 185.7, 202.5, 207.2\}$$

Calculate the mean $\bar{x}$, variance s^2, and standard deviation s.

11. Consider these data taken from a study that examines the response to ozone and sulphur dioxide among adolescents suffering from asthma. The following are measurements of forced expiratory volume (liters) for 10 subjects:

$$\{3.50, 2.60, 2.75, 2.82, 4.05, 2.25, 2.68, 3.00, 4.02, 2.85\}$$

Calculate the mean $\bar{x}$, variance s^2, and standard deviation s.

12. The percentage of ideal body weight was determined for 18 randomly selected insulin-dependent diabetics. The outcomes (%) are

107	119	99	114	120	104	124	88	114
116	101	121	152	125	100	114	95	117

Calculate the mean $\bar{x}$, variance s^2, and standard deviation s.

13. Ozone levels around Los Angeles have been measured as high as 220 parts per billion (ppb). Concentrations this high can cause the eyes to burn and are a hazard to both plant and animal life. But what about other cities? The following are data (in ppb) on the ozone level obtained in a forested area near Seattle, Washington:

160	165	170	172	161
176	163	196	162	160
162	185	167	180	168
163	161	167	173	162
169	164	179	163	178

(a) Calculate $\bar{x}$, s^2, and s.
(b) Find the median (exact).
(c) Calculate the coefficient of variation.

14. The following data are taken from a study that compares adolescents who have bulimia to healthy adolescents with similar body compositions and levels of physical activity. The following table provides measures of daily caloric intake (kcal/kg) for random samples of 23 bulimic adolescents and 15 healthy ones:

Bulimic adolescents			Healthy adolescents	
15.9	17.0	18.9	30.6	40.8
16.0	17.6	19.6	25.7	37.4
16.5	28.7	21.5	25.3	37.1
18.9	28.0	24.1	24.5	30.6
18.4	25.6	23.6	20.7	33.2
18.1	25.2	22.9	22.4	33.7
30.9	25.1	21.6	23.1	36.6
29.2	24.5		23.8	

(a) Calculate and compare the means.
(b) Calculate and compare the variances.

15. Two drugs, amantadine (A) and rimantadine (R), are being studied for use in combating the influenza virus. A single 100-mg dose is administered orally to healthy adults. The response variable is the time (in minutes) required to reach maximum concentration. Results are:

A			R		
105	123	124	221	227	280
126	108	134	261	264	238
120	112	130	250	236	240
119	132	130	230	246	283
133	136	142	253	273	516
145	156	170	256	271	
200					

(a) Calculate and compare the means.

(b) Calculate and compare the medians.

16. Using the survival data of the placebo group in Example 2.7,

(a) Compute the mean, variance, and standard deviation.

(b) How does the mean compare with the median? What does that tell us about the symmetry of the survival distribution?

(c) Can we obtain the mean and median for the 6-MP group? Why or why not?

17. Suppose the remission durations are observed for 10 patients with solid tumors. Six patients relapse at 3.0, 6.5 (ties for two), 10, 12, and 15 months; one patient is lost to follow-up at 8.4 months; and 3 patients are still in remission at the end of the study after 4.0, 5.7, and 10.0 months.

(a) Rewrite the 10 observation times in the form t or t^+ (t^+ is used for a censored case).

(b) Calculate and plot the survival time.

18. It has been assumed that life expectancy of people with mental retardation is shorter than that of the general population. However, exact estimates have not been available. To obtain these, data were collected for 99,543 persons with developmental disabilities including mental retardation, who received services from the California Department of Developmental Services between March 1984 and October 1987. Subjects were divided into three groups, and the following life table is for the most severe case ($n = 1550$); these people had severe deficits in cognitive function, were incontinent and immobile, and required tube feeding:

Age interval	Survival rate (%) at beginning
0–1	100
1–4	59
5–9	21
10–14	7
15–19	3
20–21	1

(a) Graph the survival curve.

(b) Determine the median age at death.

(c) Approximate the mean age at death, calculated as the area under the survival curve.

19. Thirty melanoma patients were studied in an effort to compare the immunotherapies BCG (Bacillus Calmette-Guerin) and c-parvum (*Corynebacterium parvum*) for their abilities to prolong remission and survival time. The following data were obtained:

BCG group (months)		C-parvum group (months)	
Remission duration	Survival time	Remission duration	Survival time
33.7+	33.7+	4.3	8.0
3.8	3.9	26.9+	26.9+
6.3	10.5	21.4+	21.4+
2.3	5.4	18.1+	18.1+
6.4	19.5	5.8	16.0+
23.8+	23.8+	3.0	6.9
1.8	7.9	11.0+	11.0+
5.5	16.9+	22.1	24.8+
16.6+	16.6+	23.0+	23.0+
33.7+	33.7+	6.8	8.3
17.1+	17.1+	10.8+	10.8+
		2.8	12.2+
		9.2	12.5+
		15.9	24.4
		4.5	7.7
		9.2	14.8+
		8.2+	8.2+
		8.2+	8.2+
		7.8+	7.8+

(a) Using remission times, calculate survival rates and plot the two curves for the two treatment groups.

(b) Repeat (a) using survival times.

20. From the following life table for male patients with localized cancer of the rectum diagnosed in two different locations:

		Location A					Location B			
Interval	n_i	d_i	c_i	n_i'	Survival rate	n_i	d_i	c_i	n_i'	Survival rate
0–1	388	167	2			749	185	10		
1–2	219	45	1			554	88	10		
2–3	173	45	1			456	55	10		
3–4	127	19	0			391	43	10		
4–5	108	17	0			338	32	14		
5–6	91	11	1			292	31	52		
6–7	79	8	0			209	20	38		
7–8	71	5	0			151	7	24		
8–9	66	6	1			120	6	25		
9–10	59	7	0			89	6	24		

(a) Complete the table with survival rates for both groups.

(b) Plot in the same graph and compare the two survival curves.

(c) Find the median survival time for each group.

3

Probability and Probability Models

3.1. PROBABILITY

3.1.1. The Certainty of Uncertainty

Even science is uncertain. Scientists are sometimes wrong. They arrive at different conclusions in many different areas: the effects of certain food ingredients or of low-level radioactivity, the role of fats in diets, and so forth. Many studies are inconclusive. For example, for decades surgeons believed that a radical mastectomy was the only treatment for breast cancer. Only recently were carefully designed clinical trials conducted to show that less drastic treatments seem equally effective.

Why is it that science is not always certain? Nature is complex and full of *unexplained biological variability*. In addition, almost all methods of observation and experiment are imperfect. Observers are subject to human bias and error. Science is a continuing story; subjects vary; measurements fluctuate. Biomedical science, in particular, contains controversy and disagreement; with the best of intentions, biomedical data—medical histories, physical examinations, interpretations of clinical tests, descriptions of symptoms and diseases—are somewhat inexact. But, most important of all, we always have to deal with incomplete information: It is either impossible, or too costly, or too time-consuming to study the entire population; we often have to rely on information gained from a *sample*—that is, a subgroup of the population under investigation. So some uncertainty almost always prevails.

Science and scientists cope with uncertainty by using the concept of *probability*. By calculating probabilities, they are able to describe what has happened and to predict what should happen in the future under similar conditions.

3.1.2. Probability

The target population of a specific research effort is the entire set of subjects at which the research is aimed. For example, in a screening for cancer in a community, the target population will consist of all persons in that community who are at risk for the disease. For one cancer site, the target population might be all women over the age of 35 years; for another site, all men over the age of 50 years.

The *probability* of an event, such as a screening test being positive, in a target population is defined as the relative frequency with which the event occurs in that target population. For example, the probability of having a disease is the disease prevalence. For another example, suppose that out of $N = 100,000$ persons of a certain target population, a total of 5,500 are positive reactors to a certain screening test; then the probability of being positive, denoted by Pr(positive), is

$$\text{Pr(Positive)} = \frac{5,500}{100,000}$$

$$= .055 \text{ or } 5.5\%$$

A probability is thus a descriptive measure for a target population with respect to a certain event. It is a number between 0 and 1 (or 0% and 100%); the larger the number, the larger the subpopulation with the event. For the case of a continuous measurement, we have the probability of being within a certain interval. For example, the probability of a serum cholesterol level between 180 and 210 (mg/100 ml) is the proportion of people in a certain target population having their cholesterol levels falling between 180 and 210 (mg/100 ml). This is measured, in the context of a histogram (see Ch. 2, Section 2.1.2), by the area of a rectangular bar for the class (180–210). Now of critical importance in the interpretation of probability is the concept of random sampling so as to associate the concept of probability with uncertainty and chance.

A sample is any subset—say n in number $(n < N)$—of the target population. Simple random sampling from the target population is sampling so that every possible sample of size n has an equal chance of selection. For simple random sampling:

1. Each individual draw is uncertain with respect to any event or characteristic under investigation (e.g., having a disease), but
2. In repeated sampling from the population, the accumulated long-run relative frequency with which the event occurs is the population relative frequency of the event.

TABLE 3.1 Cancer Screening Test Results From Example 1.4

		Test Result (X)		
		+	−	Total
Disease	+	154	225	379
(Y)	−	362	23,362	23,724
Total		516	23,587	24,103

The physical process of random sampling can be carried out as follows (or in a fashion logically equivalent to the following steps).

Step 1. A list of all N subjects in the population is obtained. Such a list is termed a *frame* of the population. The subjects are thus available to an arbitrary numbering (e.g., from 00 to $N = 99$). The frame is often based on a directory (e.g., telephone, city) or hospital records.

Step 2. A tag is prepared for each subject carrying a number $1, 2, \ldots, N$.

Step 3. The tags are placed in a receptacle, e.g., a box, and mixed thoroughly.

Step 4. A tag is drawn blindly. The number on the tag then identifies the subject from the population; this subject becomes a member of the sample.

Steps 2 to 4 can also be implemented using a table of random numbers (see Appendix A). Arbitrarily pick a 2-digit column (or 3-digit or 4-digit column if the population size is larger), and a number arbitrarily selected in that column serves to identify the subject from the population.

We can now link the concepts of probability and random sampling as follows. In the above example of cancer screening in a community of $N = 100,000$ persons, the calculated probability of .055 is interpreted as "the probability of a randomly drawn person from the target population having a positive test result is .055 or 5.5%." The rationale is as follows. On an initial draw the chosen subject may or may not be a positive reactor. However, if this process—of randomly drawing one subject at a time from the population—is repeated over and over again a large number of times, the accumulated long-run relative frequency of positive receptors in the sample will approximate .055.

3.1.3. Statistical Relationship

The data from the cancer screening test of Example 1.4 are reproduced in Table 3.1. In this design, each member of the population is characterized by two variables: the test result X and the true disease status Y. Following our above definition, the probability of a positive test result, denoted $Pr(X = +)$, is

$$Pr(X = +) = \frac{516}{24{,}103}$$

$$= .021$$

and the probability of a negative test result, denoted $Pr(X = -)$, is

$$Pr(X = -) = \frac{23{,}587}{24{,}103}$$

$$= .979$$

and similarly the probabilities of having $(Y = +)$ and not having $(Y = -)$ the disease are given by

$$Pr(Y = +) = \frac{379}{24{,}103}$$

$$= .015$$

$$Pr(Y = -) = \frac{23{,}724}{24{,}103}$$

$$= .985$$

Note that the sum of the probabilities for each variable is unity:

$$Pr(X = +) + Pr(X = -) = 1.0$$

$$Pr(Y = +) + Pr(Y = -) = 1.0$$

This is an example of the *Addition Rule* of probabilities for mutually exclusive events: one of the two events $(X = +)$ or $(X = -)$ is certain to be true for a randomly drawn individual from the population.

Furthermore, we can calculate the *joint probabilities*. These are the probabilities for two events—such as having the disease *and* having a positive test result—occurring simultaneously. With two variables, X and Y, there are four conditions of outcomes and the associated joint probabilities are

$$Pr(X = +, Y = +) = \frac{154}{24{,}103}$$

$$= .006$$

TABLE 3.2 Probabilities
Calculated for X and Y

	X		
Y	$+$	$-$	Total
$+$	.006	.009	.015
$-$	.015	.970	.985
Total	.021	.979	1.00

$$\Pr(X = +, Y = -) = \frac{362}{24,103}$$

$$= .015$$

$$\Pr(X = -, Y = +) = \frac{225}{24,103}$$

$$= .009$$

$$\Pr(X = -, Y = -) = \frac{23,362}{24,103}$$

$$= .970.$$

The second of the four joint probabilities, .015, represents the probability of a randomly drawn person from the target population having a positive test result, but being healthy (that is, a *false positive*). These joint probabilities and the above marginal probabilities separately calculated for X and Y are summarized and displayed in Table 3.2.

Observe that the four cell probabilities add to unity, i.e., one of the four events $(X = +, Y = +)$ or $(X = +, Y = -)$ or $(X = -, Y = +)$ or $(X = -, Y = -)$ is certain to be true for a randomly drawn individual from the population. Also note that the joint probabilities in each row (or column) add up to the above *marginal* or *univariate probability* at the margin of that row (or column). For example,

$$\Pr(X = +, Y = +) + \Pr(X = -, Y = +) = \Pr(Y = +)$$

$$= .015$$

We now consider a third type of probability. For example, the *sensitivity* is expressible as

$$\text{Sensitivity} = \frac{154}{379}$$

$$= .406$$

calculated for the event $(X = +)$ using the subpopulation having $(Y = +)$. That is, of the total number of 379 individuals with cancer, the proportion with a positive test result is .406 or 40.6%. This number, denoted by $\Pr(X = + \,|\, Y = +)$, is called a *conditional probability* ($Y = +$ being the condition) and is related to the other two types of probability, viz.,

$$\Pr(X = + \,|\, Y = +) = \frac{\Pr(X = +, Y = +)}{\Pr(Y = +)}$$

or

$$\Pr(X = +, Y = +) = \Pr(X = + \,|\, Y = +)\Pr(Y = +)$$

Clearly, we want to distinguish this conditional probability, $\Pr(X = + \,|\, Y = +)$, from the *marginal probability* $\Pr(X = +)$. If they are equal,

$$\Pr(X = + \,|\, Y = +) = \Pr(X = +)$$

the two events $(X = +)$ and $(Y = +)$ are said to be *independent* (because the condition $Y = +$ does not change the probability of $X = +$) and we have the *Multiplication Rule* for probabilities of independent events:

$$\Pr(X = +, Y = +) = \Pr(X = +)\Pr(Y = +)$$

If the two events are not independent, they have a statistical relationship, or we say that they are *statistically associated*. For the above screening example,

$$\Pr(X = +) = .021$$

$$\Pr(X = + \,|\, Y = +) = .406$$

clearly indicating a strong statistical relationship (because $\Pr[X = + \,|\, Y = +] \neq \Pr[X = +]$). Of course, it makes sense to have a strong statistical relationship here; otherwise the screening is useless. However, it should be emphasized that a statistical association does not necessarily mean that there is a cause and effect. Unless a relationship is so strong and so constantly repeated that the case is overwhelming, a statistical relationship, especially those observed from a sample (because the totality of population information is rarely available) is only a clue, meaning more study or confirmation is needed.

It should be noted that there are several different ways to check for the presence of a statistical relationship.

1. Calculation of odds ratio. When X and Y are independent, or not statistically associated, the odds ratio equals 1. Here we refer to the odds ratio value for the population; this value is expressed in terms of the joint probabilities as

$$\text{Odds ratio} = \frac{\Pr(X = +, Y = +)\Pr(X = -, Y = -)}{\Pr(X = +, Y = -)\Pr(X = -, Y = +)}$$

and the above example yields

$$\text{Odds ratio} = \frac{(.006)(.970)}{(.015)(.009)}$$

$$= 43.11$$

clearly indicating a statistical relationship.

2. Comparison of conditional probability and unconditional (or marginal) probability. For example, $\Pr(X = + \mid Y = +)$ versus $\Pr(X = +)$.

3. Comparison of conditional probabilities. For example, $\Pr(X = + \mid Y = +)$ versus $\Pr(X = + \mid Y = -)$. The above screening example yields

$$\Pr(X = + \mid Y = +) = .406$$

whereas

$$\Pr(X = + \mid Y = -) = \frac{362}{23,724}$$

$$= .015$$

again clearly indicating a statistical relationship. It should also be noted that we illustrate the concepts using data from a cancer screening test, but these concepts apply to any cross-classification. The primary aim is to determine whether a statistical relationship is present; Exercise 1, for example, deals with relationships between health services and race.

Finally, it is important to distinguish the two conditional probabilities, $\Pr(X = + \mid Y = +)$ and $\Pr(Y = + \mid X = +)$. In the above example

$$\Pr(X = + \mid Y = +) = \frac{154}{379}$$

$$= .406$$

whereas

$$\Pr(Y = + \mid X = +) = \frac{154}{516}$$

$$= .298$$

TABLE 3.3   Application of the Same Screening Test
to Two Different Target Populations

| | Population A | | | Population B | |
| | X | | | X | |
Y	+	−	Y	+	−
+	45,000	5,000	+	9,000	1,000
−	5,000	45,000	−	9,000	81,000

Within the context of screening test evaluation:

1. $\Pr(X = + \mid Y = +)$ and $\Pr(X = - \mid Y = -)$ are the sensitivity and specificity, respectively.
2. $\Pr(Y = + \mid X = +)$ and $\Pr(Y = - \mid X = -)$ are called the *positive predictivity* and *negative predictivity*, respectively.

With positive predictivity (or *positive predictive value*), the question is, given that the test X suggests cancer, what is the probability that, in fact, cancer is present? Rationales for these predictivities are that a test passes through several stages. Initially, the original test idea occurs to some researcher. It then must go through a developmental stage. This may have many aspects (in biochemistry, microbiology, and so forth) one of which is in biostatistics, viz., trying the test out on a pilot population. From this developmental stage, efficiency of the test is characterized by the sensitivity and specificity. An efficient test will then go through an applicational stage with an actual application of X to a target population; and here we are concerned with its predictive values. The simple example in Table 3.3 shows that, unlike sensitivity and specificity, the positive and negative predictive values depend not only on the efficiency of the test but also on the disease prevalence of the target population. In the cases presented in Table 3.3, the test is 90 percent sensitive and 90 percent specific. However,

1. Population A has a prevalence of 50%, leading to a positive predictive value of 90%.
2. Population B has a prevalence of 10%, leading to a positive predictive value of 50%.

The conclusion is clear: If a test—even a highly sensitive and highly specific one—is applied to a target population in which the disease prevalence is low (for example, population screening for a rare disease), the positive predictive value is low. (How does this relate to an important public policy: Should we conduct random testing for AIDS?)

   In the actual application of a screening test to a target population (the applicational stage), data on the disease status of individuals are not available (otherwise, screening would not be needed). However, disease prevalences are often available from national agencies and health surveys. Predictive values are then calculated from

$$\text{Positive predictivity} = \frac{(\text{Prevalence})(\text{sensitivity})}{(\text{Prevalence})(\text{sensitivity}) + (1 - \text{prevalence})(1 - \text{specificity})}$$

and

$$\text{Negative predictivity} = \frac{(1 - \text{Prevalence})(\text{specificity})}{(1 - \text{Prevalence})(\text{specificity}) + (\text{prevalence})(1 - \text{sensitivity})}$$

These formulas, called *Bayes' theorem*, allow us to calculate the predictivities without having data from the application stage. All we need are the disease prevalence (obtainable from federal health agencies) and sensitivity and specificity; these were obtained after the developmental stage. It is not too hard to prove these formulas using the addition and multiplication rules of probability; however, we choose not to present these proofs at this level. Instead of formal proofs, we now illustrate their validity using the above cancer screening example.

(i) Direct calculation of positive predictivity yields

$$\frac{9,000}{18,000} = .5$$

(ii) Use of prevalence, sensitivity, and specificity yields

$$\frac{(\text{Prevalence})(\text{sensitivity})}{(\text{Prevalence})(\text{sensitivity}) + (1 - \text{prevalence})(1 - \text{specificity})}$$

$$= \frac{(.1)(.9)}{(.1)(.9) + (1 - .1)(1 - .9)}$$

$$= .5$$

3.2. THE NORMAL DISTRIBUTION

3.2.1. Shape of the Normal Curve

The histogram in Figure 2.1 is reproduced here as Figure 3.1 (for numerical details, see Table 2.1). A close examination shows that, in general, the relative frequencies (or densities) are greatest in the vicinity of the intervals 20–29, 30–39, and 40–49, and decrease as we go toward both extremes of the range of the measurements.

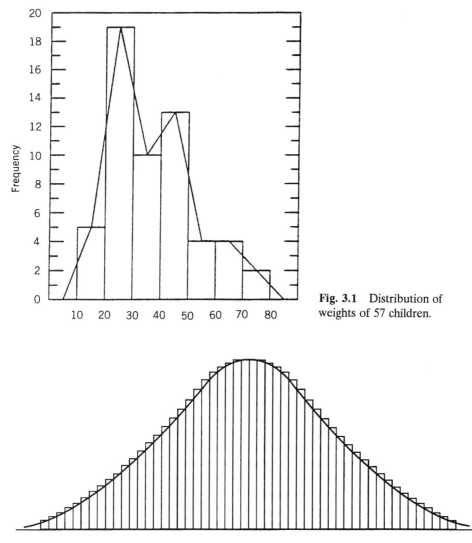

Fig. 3.1 Distribution of weights of 57 children.

Fig. 3.2 Histogram based on a large data set of weights.

Figure 3.1 shows a distribution based on a total of 57 patients; the frequency distribution consists of intervals with a width of 10 lbs. Now imagine that we increase the number of children to 50,000 and decrease the width of the intervals to .01 lb. The histogram would now look more like the one in Figure 3.2.

In Figure 3.2, the step to go from one rectangular bar to the next is very small. Finally, suppose we increase the number of children to 10 million and decrease the width of the interval to .00001 lb. You can now imagine a histogram with bars having practically no widths and the steps have all but disappeared. And if we continue to increase the size of the data set and decrease the interval width, we eventually arrive at a smooth curve that is superimposed on the histogram of Figure 3.2. You may well have already heard about the normal distribution before you took a course in statistics; many people talk about the concept. When asked to explain what it means, they (the ones who never studied statistics) describe it as a bell-shaped distribution, sort of like a handlebar moustache, similar to that in Figure 3.2. Many of them say that most distributions in nature are normal. Strictly speaking, that is false. Even more strictly speaking, they *cannot be exactly normal*. Some, such as heights of adults of a particular sex and race, are amazingly close to normal, *but never exactly*.

The normal distribution is extremely useful in statistics, but for a very different reason—not because it occurs in nature. The reason is so wonderful and amazing that it seems like magic. A couple of hundred years ago some mathematicians proved that, for samples that are "big enough," certain statistics (or summarized figures) we commonly calculate from random samples are approximately distributed as normal, even if the samples are taken from really strangely shaped distributions. There is a name given to this marvelous fact. It is called the *central limit theorem*. It is as important to statistics as the understanding of germs is to the understanding of disease. The central limit theorem tells us, for example, that values of sample mean calculated from many fairly big samples have a very nearly normal distribution. Keep in mind that "normal" is just a name for this curve; if an attribute is not distributed normally, it does not imply that it is "abnormal." Many statistics texts provide statistical procedures for finding out whether a distribution is normal; we don't give them in this text. For one thing, they are not difficult to do but are very difficult to interpret. For another, we don't think they are useful for a course at this level.

From now on, to distinguish samples from populations (a sample is a subgroup of a population), we adopt the notations given in Table 3.4. Quantities in the second column (μ, σ^2, and π) are parameters representing numerical properties of populations, μ and σ^2 for continuously measured information and π for binary information. Quantities in the first column ($\bar{x}$, s^2, and p) are statistics representing summarized information from samples. Parameters are fixed (constants) but unknown, and each statistic can be used as an estimate for the parameter listed in the same row of the above table. For example, $\bar{x}$ is used as an estimate of μ; this topic will be discussed with more details in Chapter 4. A major problem in dealing with statistics, such as $\bar{x}$ and p, is that if we repeat the example (or observational study)—even using the same sample size—values of a statistic change from sample to sample. The *Central Limit Theorem* tells us that, if sample sizes are fairly large, values of $\bar{x}$ (or p) in repeated sampling have a very nearly normal distribution. Therefore, to handle variability due to *chance*, so as to be able to declare—for example—that a certain observed difference is more than would occur by chance but is real, we first have to learn how to calculate probabilities associated with *normal curves*.

The term *normal curve*, in fact, refers not to one curve but to a family of many curves, each characterized by a mean μ and a variance σ^2. In the special case where

TABLE 3.4 Notations Used to Distinguish Between Samples
and Populations

Quantity	Notation	
	Sample	Population
Mean	$\bar{x}$ (x-bar)	μ (mu)
Variance	s^2 (s-squared)	σ^2 (sigma-squared)
Standard deviation	s	σ
Proportion	p	π (pi)

$\mu = 0$ and $\sigma^2 = 1$, we have the *standard normal curve*. For a given μ and a given σ^2, the curve is bell-shaped with the tails dipping down to the base line. In theory, the tails get closer and closer to the base line but never touch it, proceeding to infinity in either direction. In practice, we ignore that and work within practical limits.

The peak of the curve occurs at the mean μ (which for this special distribution is also median and mode), and the height of the curve at the peak depends, inversely, on the variance σ^2. Figure 3.3 shows some of these curves.

3.2.2. Areas Under the Standard Normal Curve

A variable that has a normal distribution with mean $\mu = 0$ and variance $\sigma^2 = 1$ is called the *standard normal variate* and is commonly designated by the letter Z. As with any continuous variable, probability calculations here are always concerned with finding the probability that the variable assumes any value in an interval between two specific points a and b. The probability that a continuous variable assumes a value between two points a and b is the area under the graph of the curve between a and b; the vertical axis of the graph represents the densities as defined in Chapter 2. The total area under any such curve is unity (or 100 percent), and Figure 3.4 shows the standard normal curve with some important divisions. For example, about 68% of the area is contained within ± 1, i.e.,

$$\text{Pr}(-1 < Z < 1) = .6826$$

and about 95% within ± 2,

$$\text{Pr}(-2 < Z < 2) = .9545$$

More areas under the standard normal curve have been computed and are available in tables, one of which is given in Appendix B. The entries in the table of Appendix B give the area under the standard normal curve between the mean ($Z = 0$) and a specified positive value of Z. Graphically, it is represented by the shaded region in Figure 3.5.

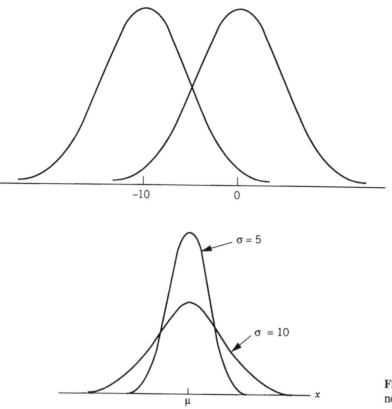

Fig. 3.3 The family of normal curves.

Using the table in Appendix B and the symmetric property of the standard normal curve, we will show how some other areas are computed.

How To Read the Table

The entries in Appendix B give the area under the standard normal curve between zero and a positive value of Z. Suppose we are interested in the area between $Z = 0$ and $Z = 1.35$ (numbers are first rounded off to two decimal places). To do this, first find the row marked with 1.3 in the left-hand column of the table, and then find the column marked with .05 in the top row of the table ($1.35 = 1.30 + .05$). Then, looking in the body of the table, we find that the "1.30 row" and the ".05 column" intersect at the value .4115. This number, .4115, is the desired area between $Z = 0$ and $Z = 1.35$. A portion of Appendix B relating to these steps is shown in Table 3.5. Another example: The area between $Z = 0$ and $Z = 1.23$ is .3907; this value is found at the intersection of the "1.2 row" and the ".03 column" of the table.

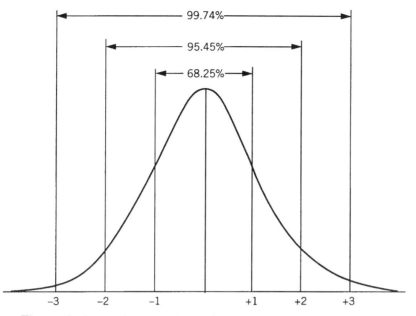

Fig. 3.4   The standard normal curve and some important divisions.

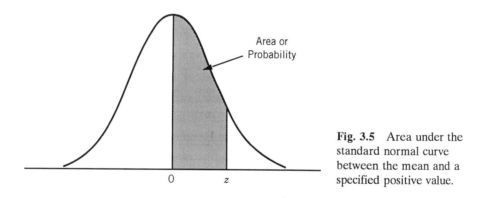

Fig. 3.5   Area under the standard normal curve between the mean and a specified positive value.

Inversely, given the area between zero and some positive value Z, we can find that value of Z. Suppose we are interested in a Z-value so that the area between zero and Z is .20. To find this Z-value, we look into the body of the table to find the tabulated area value nearest to .20, which is .2019. This number is found at the intersection of the ".5 row" and the ".03 column." Therefore, the desired Z-value is .53 (.53 = .50 + .03).

TABLE 3.5   Example From Appendix B

Z	.00	.01	.02	.03	.04	.05	etc ...
0.0							
0.1							
0.2							
$\vdots$							
1.3							.4115
$\vdots$							

Example 3.1

What is the probability of obtaining a Z-value between -1 and 1? We have

$$\Pr(-1 \leq Z \leq 1) = \Pr(-1 \leq Z \leq 0) + \Pr(0 \leq Z \leq 1)$$

$$= 2 \times \Pr(0 \leq Z \leq 1)$$

$$= (2)(.3413)$$

$$= .6826$$

which confirms the number listed in Figure 3.4. This area is shown graphically as follows:

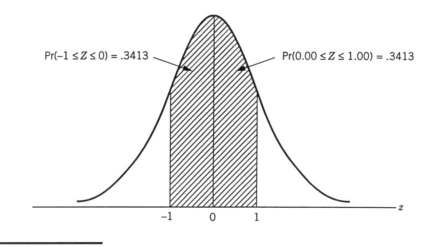

Example 3.2

What is the probability of obtaining a Z-value of at least 1.58? We have

$$\Pr(Z \geq 1.58) = .5 - \Pr(0 \leq Z \leq 1.58)$$

$$= .5 - .4429$$

$$= .0571$$

and this probability is shown below.

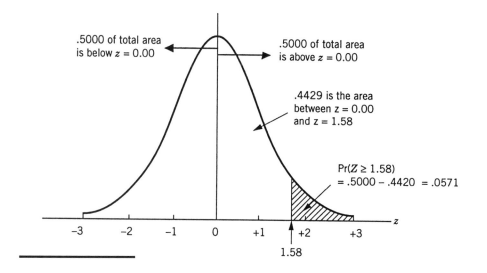

.5000 of total area
is below $z = 0.00$

.5000 of total area
is above $z = 0.00$

.4429 is the area
between $z = 0.00$
and $z = 1.58$

$\Pr(Z \geq 1.58)$
$= .5000 - .4420 = .0571$

-3   -2   -1   0   +1   +2   +3 z

1.58

Example 3.3

What is the probability of obtaining a Z-value of $-.5$ or larger? We have

$$\Pr(Z \geq -.5) = \Pr(-.5 \leq Z \leq 0) + \Pr(0 \leq Z)$$

$$= \Pr(0 \leq Z \leq .5) + \Pr(0 \leq Z)$$

$$= .1915 + .5$$

$$= .6915$$

and this probability is shown below.

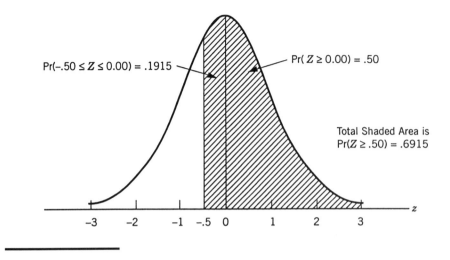

Pr(−.50 ≤ Z ≤ 0.00) = .1915

Pr(Z ≥ 0.00) = .50

Total Shaded Area is
Pr(Z ≥ .50) = .6915

Example 3.4

What is the probability of obtaining a Z-value between 1.0 and 1.58? We have

$$\Pr(1.0 \leq Z \leq 1.58) = \Pr(0 \leq Z \leq 1.58) - \Pr(0 \leq Z \leq 1.0)$$
$$= .4429 - .3413$$
$$= .1016$$

and this probability is shown below.

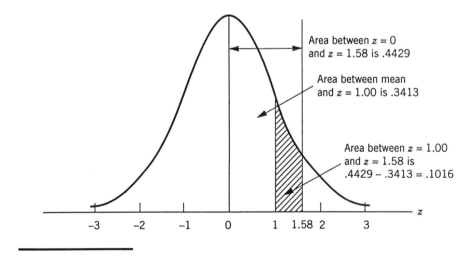

Area between z = 0
and z = 1.58 is .4429

Area between mean
and z = 1.00 is .3413

Area between z = 1.00
and z = 1.58 is
.4429 − .3413 = .1016

Example 3.5

Find a Z-value such that the probability of obtaining a larger Z-value is only .10. We have

$$Pr(Z \geq ?) = .10$$

and this is illustrated below. Scanning the table in Appendix B, we find .3997 (area between 0 and 1.28), so that

$$Pr(Z \geq 1.28) = .5 - Pr(0 \leq Z \leq 1.28)$$
$$= .5 - .3997$$
$$\cong .10$$

In terms of the question asked, there is approximately a .1 probability of obtaining a Z-value 1.28 or larger.

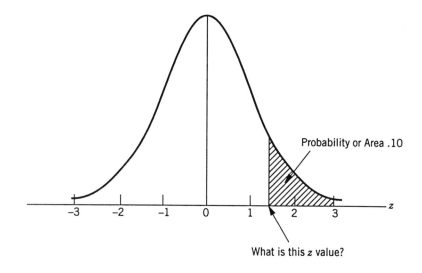

Probability or Area .10

What is this z value?

3.2.3. The Normal as a Probability Model

The reason we have been discussing the standard normal distribution so extensively with many examples is that probabilities for all normal distributions are computed using the standard normal distribution. That is, when we have a normal distribution with a given mean μ and a given standard deviation σ (or variance σ^2) for a characteristic X, we answer probability questions about the distribution by first converting to the standard normal, viz.,

$$Z = \frac{X - \mu}{\sigma}$$

Here we interpret the Z-value (or Z-score) as the number of standard deviations from the mean.

Example 3.6

If the total cholesterol values for a certain target population are approximately normally distributed with a mean of 200 (mg/100 ml) and a standard deviation of 20 (mg/100 ml), then the probability that an individual picked at random from this population will have a cholesterol value greater than 240 (mg/100 ml) is

$$\Pr(X \geq 240) = \Pr\left(\frac{X - 200}{20} \geq \frac{240 - 200}{20}\right)$$
$$= \Pr(Z \geq 2.0)$$
$$= .5 - \Pr(0 \leq Z \leq 2.0)$$
$$= .5 - .4772$$
$$= .0228 \text{ or } 2.28\%$$

Example 3.7

The following is a model for hypertension and hypotension. It is presented here as a simple illustration on the use of the normal distribution; the acceptance of the model itself is not universal.)

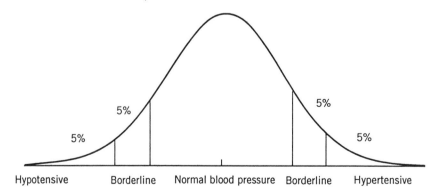

Means and standard deviation of systolic blood pressure (mm Hg) for males by age groups were collected as shown below.

Age in years	Mean	Standard deviation
16	118.4	12.17
17	121.0	12.88
18	119.8	11.95
19	121.8	14.99
20–24	123.9	13.74
25–29	125.1	12.58
30–34	126.1	13.61
35–39	127.1	14.20
40–44	129.0	15.07
45–54	132.3	18.11
55–64	139.8	19.99

From this table, and using the table in Appendix B, systolic blood pressure limits for each group can be calculated as follows:

Age	Hypotension if below	Lowest healthy	Highest healthy	Hypertension if above
16	98.34	102.80	134.00	138.46
17	99.77	104.49	137.51	142.23
18	100.11	104.48	135.12	139.49
19	97.10	102.58	141.02	146.50
20–24	?	?	?	?
25–29	?	?	?	?
30–34	103.67	108.65	143.55	148.53
35–39	103.70	108.90	145.30	150.50
40–44	104.16	109.68	148.32	153.84
45–54	102.47	109.09	155.41	162.03
55–64	106.91	114.22	165.38	172.74

For example, the highest healthy limit for the 20–24 years group is obtained as follows:

$$\Pr(X \geq ?) = .10$$

$$= \Pr\left(\frac{X - 123.9}{13.74} \geq \frac{? - 123.9}{13.74}\right)$$

and from Example 3.5, we have

$$1.28 = \frac{? - 123.9}{13.74}$$

leading to

$$? = 123.9 + (1.28)(13.74)$$
$$= 141.49$$

3.3. SOME OTHER DISTRIBUTIONS

3.3.1. The Binomial Distribution

In Chapter 1, we discussed cases with dichotomous outcomes such as Male–Female, Survived–Not survived, Infected–Not infected, or White–Non-White. We have seen that such data can be summarized into proportions, rates, and ratios. In this section, we are concerned with the probability of a compound event: the occurrence of x outcomes ($0 \leq x \leq n$) in n trials, called a *binomial probability*. For example, if a certain drug is known to cause a side effect 10% of the time and if five patients are given this drug, then what is the probability that four or more experience the side effect?

Let S denote a side effect outcome and N an outcome without side effects. The process of determining the chance of x S's in n trials consists of listing all the possible mutually exclusive outcomes, calculating the probability of each outcome using the multiplication rule (where the trials are assumed to be independent), and then combining the probability of all those outcomes that are compatible with the desired results using the addition rule. With five patients there are 32 mutually exclusive outcomes, as shown in Table 3.6.

Since the five patient outcomes are independent, the multiplication rule produces the probabilities shown for each outcome. For example,

- The probability of obtaining an outcome with 4 S's and 1 N is

$$(.1)(.1)(.1)(.1)(1 - .1) = (.1)^4(.9)$$

- The probability of obtaining all 5 S's is

$$(.1)(.1)(.1)(.1)(.1) = (.1)^5$$

Since the event "all five with side effects" corresponds to only one of the 32 outcomes above and the event "four with side effects and one without" pertains to 5 of

TABLE 3.6 Possible Outcomes in Five Independent Trials and Their
Associated Probabilities

| | Outcome | | | | | No. of |
First patient	Second patient	Third patient	Fourth patient	Fifth patient	Probability	patients having side effects
S	S	S	S	S	$(.1)^5$	→ 5
S	S	S	S	N	$(.1)^4(.9)$	→ 4
S	S	S	N	S	$(.1)^4(.9)$	→ 4
S	S	S	N	N	$(.1)^3(.9)^2$	3
S	S	N	S	S	$(.1)^4(.9)$	→ 4
S	S	N	S	N	$(.1)^3(.9)^2$	3
S	S	N	N	S	$(.1)^3(.9)^2$	3
S	S	N	N	N	$(.1)^2(.9)^3$	2
S	N	S	S	S	$(.1)^4(.9)$	→ 4
S	N	S	S	N	$(.1)^3(.9)^2$	3
S	N	S	N	S	$(.1)^3(.9)^2$	3
S	N	S	N	N	$(.1)^2(.9)^3$	2
S	N	N	S	S	$(.1)^3(.9)^2$	3
S	N	N	S	N	$(.1)^2(.9)^3$	2
S	N	N	N	S	$(.1)^2(.9)^3$	2
S	N	N	N	N	$(.1)(.9)^4$	1
N	S	S	S	S	$(.1)^4(.9)$	→ 4
N	S	S	S	N	$(.1)^3(.9)^2$	3
N	S	S	N	S	$(.1)^3(.9)^2$	3
N	S	S	N	N	$(.1)^2(.9)^3$	2
N	S	N	S	S	$(.1)^3(.9)^2$	3
N	S	N	S	N	$(.1)^2(.9)^3$	2
N	S	N	N	S	$(.1)^2(.9)^3$	2
N	S	N	N	N	$(.1)(.9)^4$	1
N	N	S	S	S	$(.1)^3(.9)^2$	3
N	N	S	S	N	$(.1)^2(.9)^3$	2
N	N	S	N	S	$(.1)^2(.9)^3$	2
N	N	S	N	N	$(.1)(.9)^4$	1
N	N	N	S	S	$(.1)^2(.9)^3$	2
N	N	N	S	N	$(.1)(.9)^4$	1
N	N	N	N	S	$(.1)(.9)^4$	1
N	N	N	N	N	$(.9)^5$	0

the 32 outcomes, each with probability $(.1)^4(.9)$, the addition rule yields a probability

$$(.1)^5 + (5)(.1)^4(.9) = .00046$$

for the compound event that "four or more have side effects". In summary, *there are formulas to calculate the exact probability of the compound events involving x positive outcomes from n trials; however, these are not simple* and not useful for a course at this level.

When the number of trials n is from moderate to large ($n > 25$, say), we can answer probability questions by first converting to the standard normal score

$$Z = \frac{X - n\pi}{\sqrt{n\pi(1-\pi)}}$$

where π is the probability of having a positive outcome from a single trial. In other words, we approximate the binomial distribution by a normal distribution with mean $\mu = n\pi$ and variance $\sigma^2 = n\pi(1-\pi)$. For example, for $\pi = .1$ and $n = 30$, we have

$$\mu = (30)(.1)$$

$$= 3$$

$$\sigma^2 = (30)(.1)(.9)$$

$$= 2.7$$

so that

$$\Pr(X \geq 7) \cong \Pr\left(Z \geq \frac{7-3}{\sqrt{2.7}}\right)$$

$$= \Pr(Z \geq 2.43)$$

$$= .0075$$

In other words, if the true probability for having the side effect is 10%, then the probability of having 7 or more of 30 patients with the side effect is less than 1% ($= .0075$).

3.3.2. The Poisson Distribution

The next discrete distribution that we consider is the Poisson distribution, named after a French mathematician. This distribution has been used extensively in health science to model the distribution of the number of occurrences X of some random event in an interval of time or space, or some volume of matter. For example, a hospital administrator has been studying daily emergency admissions over a period of several months and has found that admissions have averaged three per day. He or she is then interested in finding the probability that no emergency admissions will occur on a particular day. Similar to the case of binomial probabilities, there are formulas to calculate exact Poisson probabilities; however, these are not simple and are beyond the scope of this book. It turns out, interestingly enough, that for a Poisson distribution the variance is equal to the mean. If we denote by θ the average occurrence for a Poisson distribution, we can answer probability questions by first connecting the number of occurrences X to the standard normal score, provided

$\theta \geq 10$:

$$Z = \frac{x - \theta}{\sqrt{\theta}}$$

In other words, we can approximate a Poisson distribution by a normal distribution with mean θ if θ is at least 10.

Here is another example involving the Poisson distribution. The infant mortality rate (IMR) is defined as

$$IMR = d/N$$

for a certain target population during a given year where d is the number of deaths during the first year of life and N is the total number of live births. In the studies of IMRs, N is conventionally assumed as fixed and d to follow a Poisson distribution.

Example 3.8

For the year 1981, we have the following data for the New England states (Connecticut, Maine, Massachusetts, New Hampshire, Rhode Island, and Vermont):

$$d = 1,585$$

$$N = 164,200$$

For the same year, the national IMR was 11.9 (per 1,000 live births). If we apply national IMR to the New England states, then we would have

$$\theta = (11.9)(164.2)$$

$$\cong 1,954 \text{ infant deaths}$$

Then the event of having as few as 1,585 infant deaths would occur with a probability

$$Pr(d \leq 1,585) = Pr\left(Z \leq \frac{1,585 - 1,954}{\sqrt{1,954}}\right)$$

$$= Pr(Z \leq -8.35)$$

$$\cong 0$$

The conclusion is clear: Either we observed an extremely improbable event, or the infant mortality of New England states is lower than the national average. The observed rate for New England states was 9.7 deaths per 1,000 live births.

3.3.3. Other Statistical Tables

In addition to the normal distribution (Appendix B), topics introduced in subsequent chapters will involve two other continuous distributions:

- the t distribution (Appendix C)
- the chi-square distribution (Appendix D)

The t distribution is similar to the standard normal distribution in that it is unimodal, bell-shaped, and symmetric, and it extends infinitely in either direction. This is a family of curves, each indexed by a number called the *degrees of freedom* (df). Degrees of freedom measure the quantity of information available in a data set that can be used for estimating the population variance σ^2. Although t curves have more variance than the standard normal, the area under each curve is still equal to unity (or 100%). Areas under a curve from the right tail, shown by the shaded region, are listed in Appendix C; the t distribution for infinite degrees of freedom is precisely equal to the standard normal distribution. This equality is readily seen by examining the column marked with, say, Area = .025. The last row (infinite df) shows a value of 1.96 that can be verified using the table in Appendix B.

Unlike the normal and the t distributions, the chi-square is concerned with non-negative attributes and will be used only for certain "tests" in Chapter 6.

EXERCISES

1. Although cervical cancer is not a leading cause of death among American women, it has been suggested that virtually all such deaths are preventable (5,166 American women died from cervical cancer in 1977). In an effort to find out who is being or not being screened for cervical cancer (Pap testing), the following data were collected from a certain community:

Pap test	White	Black	
No	5,244	785	6,029
Yes	25,117	2,348	27,465
	30,361	3,133	33,494

Is there a statistical relationship here? (Try a few different methods: calculation of odds ratio, comparison of conditional and unconditional probabilities, and comparison of conditional probabilities.)

2. In a study of intraobserver variability in assessing cervical smears, 3,325 slides were screened for the presence or absence of abnormal squamous cells. Each slide was screened by a particular observer and then re-screened 6 months later by the same observer. The results are as follows:

First	Second screening		
screening	Present	Absent	Total
Present	1,763	489	2,252
Absent	403	670	1,073
Total	2,166	1,159	3,325

Is there a statistical relationship between first screening and second screening? (Try a few different methods as in the previous exercise.)

3. From the above intraobserver variability study, find:
    (a) The probability that abnormal squamous cells were found to be absent in both screenings.
    (b) The probability of an absence in the second screening given that abnormal cells were found in the first screening.
    (c) The probability of an abnormal presence in the second screening given that no abnormal cells were found in the first screening.
    (d) The probability that the screenings disagree.
4. Given the screening test of Example 1.4, where

$$\text{Sensitivity} = .406$$

$$\text{Specificity} = .985$$

calculate the positive predictive values when the test is applied to the following populations:

Population A: 80% prevalence

Population B: 25% prevalence

5. Consider the following data on the use of x-ray as a screening test for tuberculosis.

X-ray	Tuberculosis	
	No	Yes
Negative	1,739	8
Positive	51	22
Total	1,790	30

(a) Calculate the sensitivity and specificity.
(b) Find the disease prevalence.
(c) Calculate the positive predictive value both directly and indirectly (using Bayes' theorem), as in Section 3.1.3.

6. From the sensitivity and specificity of x-rays found in Exercise 5, compute the positive predictive values corresponding to these prevalences: .2, .4, .6, .7, .8, and .9. Can we find a prevalence when the positive predictive value is pre-set at .80 or 80%?

7. Refer to the standard normal distribution. What is the probability of obtaining a Z-value of
   (a) at least 1.25?
   (b) at least $-.84$?

8. Refer to the standard normal distribution. What is the probability of obtaining a Z-value
   (a) between -1.96 and 1.96?
   (b) between 1.22 and 1.85?
   (c) between $-.84$ and 1.28?

9. Refer to the standard normal distribution. What is the probability of obtaining Z-value
   (a) less than 1.72?
   (b) less than -1.25?

10. Refer to the standard normal distribution. Find a Z-value such that the probability of obtaining a larger Z-value is
    (a) .05;
    (b) .025;
    (c) .20.

11. Complete the table in Example 3.7 at the question marks.

12. Medical research has concluded that individuals experience a common cold roughly two times per year. Assume that the time between colds is normally distributed with a mean of 160 days and a standard deviation of 40 days.
    (a) What is the probability of going 200 or more days between colds? Of going 365 or more days?
    (b) What is the probability of getting a cold within 80 days of a previous cold?

13. Assume that the test scores for a large class are normally distributed with a mean of 74 and a standard deviation of 10.
    (a) Suppose you receive a score of 88. What percent of the class received scores higher than this?
    (b) Suppose the teacher wants to limit the number of A grades in the class to no more than 20%. What would be the lowest score for an A?

14. Intelligence test scores, referred to as *intelligence quotient* or *IQ* scores, are based on characteristics such as verbal skills, abstract reasoning power, numerical ability, and spatial visualization. If plotted on a graph, the distribution of IQ scores approximates a normal curve with a mean of about 100. An IQ score above 115 is considered superior. Studies of "intellectually gifted" children have generally defined the lower limit of their IQ scores at 140; approximately 1% of the population have IQ scores above this limit.

(a) Find the standard deviation of this distribution.

(b) What percent are in the "superior" range of 115 or above?

(c) What percent of the population have IQ scores of 70 or below?

15. IQ scores for college graduates are normally distributed with a mean of 120 (as compared to 100 for the general population) with a standard deviation of 12. What is the probability of randomly selecting a graduate student with an IQ score

(a) Between 110 and 130?

(b) Above 140?

(c) Below 100?

16. Suppose it is known that the probability of recovery for a certain disease is .4. If 35 people are stricken with the disease, what is the probability that

(a) 25 or more will recover?

(b) Fewer than 5 will recover?

(Of course, use the normal approximation.)

17. A study found that for 60% of the couples who have been married 10 years or less, both spouses work. A sample of 30 couples who have been married 10 years or less are selected from marital records available at a local courthouse. We are interested in the number of couples in this sample in which both spouses work. What is the probability that this number is

(a) 20 or more?

(b) 25 or more?

(c) 10 or fewer?

18. Many samples of water, all the same size, are taken from a river suspected of having been polluted by irresponsible operators at a sewage treatment plant. The number of coliform organisms in each sample was counted; the average number of organisms per sample was 15. Assuming the number of organisms to be Poisson-distributed, find the probability that

(a) The next sample will contain at least 20 organisms.

(b) The next sample will contain no more than 5 organisms.

19. For the year 1981 (see Example 3.8), we also have the following data for the South Atlantic states (Delaware, Florida, Georgia, Maryland, North and South Carolina, Virginia, West Virginia, and the District of Columbia):

$$d = 7,643 \text{ infant deaths}$$

$$N = 550,300 \text{ live births}$$

Find the infant mortality rate and compare it with the national average using the method of Example 3.8.

20. For the t curve with 20 df, find the areas

(a) To the left of 2.086 and of 2.845

(b) To the right of 1.725 and of 2.528

(c) Beyond ± 2.086 and beyond ± 2.845

CHAPTER

4

Confidence Estimation

This chapter deals with a statistical procedure called *estimation*. It is extremely useful, one of the most useful procedures of statistics. The word "estimate" actually has a language problem, the opposite of the language problem of statistical "tests" (the topic of Chapter 5). The colloquial meaning of the word *test* makes one new to the field of statistics think that statistical tests are especially objective, no-nonsense procedures that reveal the truth. Conversely, the colloquial meaning of the word "estimate" is that of guessing, perhaps off the top of the head and uninformed, not to be taken too seriously. It is used by car body repair shops, who "estimate" how much it will cost to fix your car after an accident. The "estimate" in that case is actually a bid of a for-profit business establishment seeking your trade. This language problem was created by statisticians, who wanted to convey meaning without inventing new jargon. The word "estimation" isn't that bad a choice, once you understand what statisticians mean by it. As you learn it you'll see the connection between the colloquial meaning and the technical meaning. First, let's make it clear that statistical estimation is no less objective than any other formal statistical procedure. Statistical estimation requires calculations and tables just as statistical testing does. Estimation, however, has a different purpose. It uses data as a basis for calculations of estimates of quantities of interest, as you might expect. What it adds, however, is something extremely important. This addition is what separates formal statistical estimation from ordinary guessing. It is the determination of the *amount of uncertainty* in the estimate. How often have you heard of someone making a guess and then giving you a number measuring the uncertainty of the guess? That's what statistical estimation does. It gives you the best guess and then tells you how shaky the guess is, in quite precise terms.

Certain media, sophisticated newspapers in particular, are starting to educate the public about statistical estimation. They do it when they report the results of polls. They say things like, "74% of the voters disagree with the governor's budget proposal," and then go on to say that the sampling error is plus or minus 3%. What they are saying is that whoever conducted the poll is claiming to have polled about 1,000 people chosen at random and that statistical estimation theory tells us to be 95% certain that if *all* the voters were polled their disagreement percentage would be discovered to be within 3% of 74%. In other words, it's very unlikely that the 74% is off the mark by more than 3%; the truth is almost certainly between 71% and 77%. (If you study more advanced statistical theory you'll learn the subtleties of statistical certainty and the strict interpretation of confidence intervals, but we can keep it simple and heuristic for now and still get the main idea.)

4.1. BASIC CONCEPTS

A class of measurements or a characteristic on which individual observations or measurements are made is called a *variable* or *random variable*. The value of a random variable varies from subject to subject; examples include weight, height, blood pressure, or the presence or absence of a certain habit or practice, such as smoking or using drugs. The distribution of a random variable is often assumed to belong to a certain family of distributions such as binomial, Poisson, or normal. This assumed family of distributions is specified or indexed by one or several parameters such as a population mean μ or a population proportion π. It is usually impossible, too costly, or too time-consuming to obtain the entire population data on any variable in order to learn about a parameter involved in its distribution. Decisions in health science are thus often made using a small sample of a population. The problem for a decision maker is to decide on the basis of data the estimated value of a parameter, such as the population mean, as well as to provide certain ideas concerning errors associated with that estimate.

4.1.1. Statistics as Variables

A parameter is a numerical property of a population; examples include population mean μ and population proportion π. The corresponding quantity obtained from a sample is called a *statistic;* examples of statistics include the sample mean $\bar{x}$ and sample proportion p. Statistics help us to draw inferences or conclusions about population parameters. After a sample has already been obtained, the value of a statistic—for example, the sample mean $\bar{x}$—is known and fixed; however, if we take a different sample we almost certainly have a different numerical value for that same statistic. In this repeated sampling context, a statistic is looked upon as a variable that takes different values from sample to sample.

4.1.2. Sampling Distributions

The distribution of values of a statistic obtained from repeated samples of the same size from a given population is called the *sampling distribution* of that statistic.

Example 4.1

Consider a population consisting of six subjects (this small size is impractical, but we need something small enough to use as an illustration here); the following table gives the subject names (for identification) and values of a variable under investigation (for example, 1 for a smoker and 0 for a non-smoker):

Subject	Value
A	1
B	1
C	1
D	0
E	0
F	0

In this case the population mean μ (also population proportion π for this very special dichotomous variable) is .5 ($= 3/6$). We now consider all possible samples, without replacement, of size 3; none or some or all subjects in each sample have value "1," the remaining, "0." The following table represents the sampling distribution of the sample mean:

Samples	No. of samples	Value of sample mean $\bar{x}$
(D, E, F)	1	0
(A, D, E), (A, D, F), (A, E, F) (B, D, E), (B, D, F), (B, E, F) (C, D, E), (C, D, F), (C, E, F)	9	1/3
(A, B, D), (A, B, E), (A, B, F) (A, C, D), (A, C, E), (A, C, F) (B, C, D), (B, C, E), (B, C, F)	9	2/3
(A, B, C)	1	1
Total	20	

This sampling distribution gives us a few interesting properties:

(i) Its mean, i.e., the mean of all possible sample means, is

$$\frac{(1)(0) + (9)(1/3) + (9)(2/3) + (1)(1)}{20} = .5$$

which is the same as the mean of the original distribution. Because of this we say that the sample mean (sample proportion) is an *unbiased estimator* for the population mean (population proportion). In other words, if we use the sample mean (sample proportion) to estimate the population mean (population proportion), we are correct on the average.

(ii) If we form a bar graph for this sampling distribution,

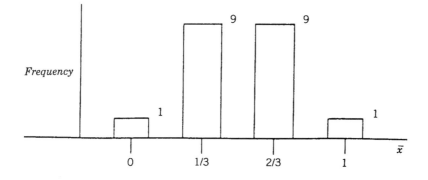

it shows a shape somewhat similar to that of a symmetric, bell-shaped normal curve. This resemblance is much clearer with real populations and larger sample sizes.

We now consider the same population and all possible samples of size $n = 4$. The following table represents the new sampling distribution:

Samples	No. of samples	Value of sample mean $\bar{x}$
(A, D, E, F), (B, D, E, F), (C, D, E, F)	3	.25
(A, B, D, E), (A, B, D, F), (A, B, E, F) (A, C, D, E), (A, C, D, F), (A, C, E, F) (B, C, D, E), (B, C, D, F), (B, C, E, F)	9	.50
(A, B, C, D), (A, B, C, E), (A, B, C, F)	3	.75
Total	15	

It can be seen that we have a different sampling distribution because the sample size is different. However, we still have both above-mentioned properties:

(i) Unbiasedness of the sample mean

$$\frac{(3)(.25) + (9)(.50) + (3)(.75)}{15} = .5$$

(ii) Normal shape of the sampling distribution (bar graph)

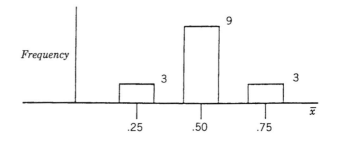

In addition, we can see that

(iii) The variance of the new distribution is smaller. The two faraway values of $\bar{x}$, 0 and 1, are no longer possible; new values—.25 and .75—are closer to the mean .5 and the majority (nine samples) have values that are right at the sampling distribution mean. The major reason for this is that the new sampling distribution is associated with a larger sample size, $n = 4$ as compared with $n = 3$.

4.1.3. Introduction to Confidence Estimation

Statistical inference is the procedure whereby inferences about a population are made on the basis of the results obtained from a sample drawn from that population.

Professionals in health science are often interested in a parameter of a certain population. For example, a health professional may be interested in knowing what proportion of a certain type of individual, treated with a particular drug, suffers undesirable side effects. The process of estimation entails calculating, from the data of a sample, some statistic that is offered as an estimate of the corresponding parameter of the population from which the sample was drawn.

A point estimate is a single numerical value used to estimate the corresponding population parameter. For example, the sample mean is a point estimate for the population mean and the sample proportion is a point estimate for the population proportion. However, having access to the data of a sample and a knowledge of statistical theory, we can do more than just provide a point estimate. The sampling distribution of a statistic—if available—would provide information on

(i) Biasedness/unbiasedness—several statistics, such as $\bar{x}$, p, and s^2 are unbiased
(ii) Variance

Variance is important; a small variance for a sampling distribution indicates that most possible values for the statistic are close to each other so that a particular value is more likely to be reproduced. In other words, the variance of a sampling distribution of a statistic can be used as a measure of precision or reproducibility of that statistic; the smaller this quantity, the better the statistic as an estimate of the corresponding parameter. The square root of this variance is called the *standard error* of the statistic; for example, we will have the standard error of the sample mean, or $SE(\bar{x})$, standard error of the sample proportion, $SE(p)$, and so on. In the next few sections, we will introduce a process whereby the point estimate and its standard error are combined to form an interval estimate or a *confidence interval*. A confidence interval consists of two numerical values defining an interval that, with a specified degree of confidence, we feel includes the parameter being estimated.

4.2. ESTIMATION OF MEANS

The results in Example 4.1 are not coincidences but are examples of the characteristics of sampling distributions in general. The key tool here is the *central limit theorem*, introduced in Section 3.2.1, which may be summarized as follows:

Given any population with mean μ and variance σ^2, the sampling distribution of $\bar{x}$ will be approximately normal with mean μ and variance σ^2/n when the sample size n is large (of course, the larger the sample size, the better the approximation; in practice, an $n = 25$ or more could be considered adequately large). This means

$$\mu_{\bar{x}} = \mu$$

$$\sigma_{\bar{x}}^2 = \sigma^2/n$$

The following example will show how good $\bar{x}$ is as an estimate for the population μ even if the sample size is as small as 25 (of course, it is used only as an illustration; in practice, μ and σ^2 are unknown).

Example 4.2

Birth weights obtained from deliveries over a long period of time at Boston City Hospital show a mean μ of 112 ounces and a standard deviation σ of 20.6 ounces. Let us suppose we want to compute the probability that the mean birth weight from a sample of 25 infants will fall between 107 and 117 ounces. The Central Limit Theorem is applied, and it indicates that $\bar{x}$ follows a normal distribution with mean

$$\mu_{\bar{x}} = 112$$

and variance

$$\sigma_{\bar{x}}^2 = (20.6)^2/25$$

or standard deviation

$$\sigma_{\bar{x}} = 4.12$$

It follows that

$$\Pr(107 \leq \bar{x} \leq 117) = \Pr\left(\frac{107 - 112}{4.12} \leq z \leq \frac{117 - 112}{4.12}\right)$$

$$= \Pr(-1.21 \leq z \leq 1.21)$$

$$= (2)(.3869)$$

$$= .7738$$

In other words, if we use the mean of a sample of size $n = 25$ to estimate the population mean, about 80% of the time we are correct within 5 ounces; this figure would be 98.5% if the sample size is 100.

4.2.1. Confidence Intervals for a Mean

Similar to what was done in Example 4.2, we can write, for example,

$$\Pr\left[-1.96 \leq \frac{\bar{x} - \mu}{\sigma/\sqrt{n}} \leq 1.96\right] = (2)(.475)$$

$$= .95$$

This statement is a consequence of the Central Limit Theorem, which indicates that, for a large sample size n, $\bar{x}$ is a random variable (in the context of repeated sampling) with a normal sampling distribution in which

$$\mu_{\bar{x}} = \mu$$

$$\sigma_{\bar{x}}^2 = \sigma^2/n$$

The quantity inside the square bracket of the above equation is equivalent to

$$\bar{x} - 1.96\sigma/\sqrt{n} \le \mu \le \bar{x} + 1.96\sigma/\sqrt{n}$$

all we need to do now is to select a random sample, calculate the numerical value of $\bar{x}$ and its standard error with σ replaced by sample variance s, $s/\sqrt{n}$, and substitute these values to form the end points of the interval,

$$\bar{x} \pm 1.96s/\sqrt{n}$$

In a specific numerical case this will produce two numbers,

$$a = \bar{x} - 1.96s/\sqrt{n}$$

and

$$b = \bar{x} + 1.96s/\sqrt{n}$$

and we have the interval

$$a \le \mu \le b$$

But here we run into a problem. We are sampling from a fixed population. We are examining values of a random variable obtained by selecting a random sample from that fixed population. The random variable has a distribution with mean μ that we wish to estimate. Since the population and the distribution of the random variable we are investigating are fixed, it follows that the parameter μ is fixed. The quantities μ, a and b are all fixed (after the sample has been obtained); then we cannot assert the probability that μ lies between a and b is .95. In fact, either μ lies in (a,b) or it does not, and it is not correct to assign a probability to the statement (even the truth remains unknown).

The difficulty here arises at the point of substitution of the observed numerical values for $\bar{x}$ and its standard error. The random variation in $\bar{x}$ is variation from sample to sample in the context of repeated sampling. When we substitute $\bar{x}$ and its standard error $s/\sqrt{n}$ by their numerical values resulting in interval (a,b), it is understood that the repeated sampling process could produce many different intervals of the same form

$$\bar{x} \pm (1.96)\text{SE}(\bar{x})$$

About 95% of these intervals would actually include μ. Since we have only one of these possible intervals, that is, the interval (a,b) from our sample, we say we are 95% confident that μ lies between these limits. The interval (a,b) is called a

TABLE 4.1 Frequently Used Confidence Intervals

Degree of confidence (%)	Coefficient
99	2.576
→ 95	1.960
90	1.645
80	1.282

95% percent confidence interval for μ and the figure "95" is called the *degree of confidence* or *confidence level*.

In forming confidence intervals, the degree of confidence is determined by the investigator of a research project. Different investigators may prefer different confidence intervals; the coefficient to be multiplied with the standard error of the mean should be determined accordingly. A few typical choices are shown in Table 4.1; 95% is the most conventional. Finally, it should be noted that since the standard error is

$$SE(\overline{x}) = s/\sqrt{n}$$

the width of a confidence interval becomes narrower as sample size increases, and the above process is applicable only to large samples ($n > 25$, say). The next section will show how to handle smaller samples (there is nothing magic about "25"; see note at the end of Section 4.2.2).

_____

Example 4.3

For the data on percentage saturation of bile for 31 male patients of Example 2.4:

40, 86, 111, 86, 106, 66, 123, 90, 112, 52, 88, 137, 88, 88, 65, 79, 87, 56, 110, 106, 110, 78, 80, 47, 74, 58, 88, 73, 118, 67, 57

we have

$$n = 31$$

$$\overline{x} = 84.65$$

$$s = 24.00$$

TABLE 4.2 t-Coefficients for Use With 95 Percent Confidence Intervals

df	t-coefficient (percentile)
5	2.571
10	2.228
15	2.131
20	2.086
24	2.064
$\rightarrow \infty$	1.960

leading to a standard error

$$SE(\bar{x}) = 24.00/\sqrt{31}$$

$$= 4.31$$

and a 95% confidence interval for the population mean

$$84.65 \pm (1.96)(4.31) = (76.2, 93.1)$$

(The resulting interval is wide due to a large standard deviation as observed from the sample $s = 24.0$, reflecting heterogeneity of sample subjects.)

4.2.2. Uses of Small Samples

The procedure in the previous section for confidence intervals is applicable only to large samples ($n > 25$). For smaller samples, the results are still valid if the population variance σ^2 is known and the standard error is expressed as $\sigma/\sqrt{n}$. However, σ^2 is almost always unknown. When σ is unknown, we can estimate it by s, but the procedure has to be modified by changing the coefficient to be multiplied by the standard error; how much larger the coefficient is depends on how much information we have in estimating σ (by s), that is, the sample size n.

Therefore, instead of taking coefficients from the standard normal distribution table (numbers 2.576, 1.960, 1.645, and 1.282), we will use corresponding numbers from the t-curves where the quantity of information is indexed by the degree of freedom (df $= n - 1$). The figures are listed in Appendix C; the column to read is the one with the correct normal coefficient on the bottom row (marked with df $= \infty$). Some t-coefficients associated with 95 percent confidence are given in Table 4.2 for various degrees of freedom. For better results, it is a good practice always to use the t-table regardless of sample size, because coefficients such as 1.96 are only for very large sample sizes.

Example 4.4

In an attempt to assess the physical condition of joggers, a sample of $n = 25$ joggers was selected and maximum volume oxygen (VO_2) uptake was measured with the following results:

$$\bar{x} = 47.5 \text{ ml/kg}$$

$$s = 4.8 \text{ ml/kg}$$

From Appendix C we find that the t-coefficient with 24 df for use with a 95% confidence interval is 2.064. We have

$$SE(\bar{x}) = 4.8/\sqrt{25}$$

$$= .96$$

and a 95% confidence interval for the population mean μ (this is the population of joggers' VO_2 uptake) is

$$47.5 \pm (2.064)(.96) = (45.5, 49.5)$$

The previous two sections deal with the determination of confidence intervals for a population mean. The rest of this chapter covers other parameters: the difference of two population means, one population proportion, the difference of two population proportions, and the odds ratio. We could present each topic in full detail as in the case of one population mean; however, this is repetitive, so we choose to shorten the description and retain only the mechanical part. Interpretations are similar to those of the previous two sections.

4.2.3. Difference of Means

For the comparison of means, for example, in efforts to determine the effect of a risk factor or an intervention, we distinguish between two situations. Sometimes we take two independent samples, but in many investigations the experimental group is also used as its own control. The latter technique—matched design or before-and-after intervention—often generates quite appropriate comparisons because variability due to extraneous factors is reduced (it is not unusual for extraneous factors to account for many of the differences between means obtained from two independent samples).

Case 1: Matched Samples

Data from matched or before-and-after experiments should not be considered as coming from two independent samples. The procedure is to reduce the data to a one-sample problem by computing before-and-after (or control-and-case) differences for each subject (or pairs of matched subjects). By doing this with paired observations, we get a set of differences that can be handled as a single-sample problem.

Example 4.5

The systolic blood pressures of 12 women between the ages of 20 and 35 years were measured before and after administration of a newly developed oral contraceptive. Given the data in the following table on systolic blood pressure (mmHg), we have from the column of differences, the d_i's,

$$n = 12$$

$$\sum d_i = 31$$

$$\sum d_i^2 = 185$$

Subject	Before	After	After–before difference, d_i	d_i^2
1	122	127	5	25
2	126	128	2	4
3	132	140	8	64
4	120	119	−1	1
5	142	145	3	9
6	130	130	0	0
7	142	148	6	36
8	137	135	−2	4
9	128	129	1	1
10	132	137	5	25
11	128	128	0	0
12	129	133	4	16

leading to

$$\bar{d} = \text{Average difference}$$

$$= 31/12$$

$$= 2.58 \text{ mmHg}$$

$$s^2 = \frac{185 - (31)^2/12}{11}$$

$$= 9.54$$

$$s = 3.09$$

$$SE(\bar{d}) = 3.09/\sqrt{12}$$

$$= .89$$

With a degree of confidence of .95 the t-coefficient from Appendix C is 2.201 so that a 95% confidence interval for the mean difference is

$$2.58 \pm (2.201)(.89) = (.62, 4.54)$$

That means the "after" mean is larger than the "before" mean, an increase of between .62 and 4.54.

Case 2: Independent Samples

For the case of two independent samples the procedure is as follows:

1. Data are summarized separately to obtain

$$Sample\ 1: \quad n_1, \bar{x}_1, s_1^2$$

$$Sample\ 2: \quad n_2, \bar{x}_2, s_2^2$$

2. Standard error of the difference of means is given by

$$SE(\bar{x}_1 - \bar{x}_2) = \sqrt{\frac{s_1^2}{n_1} + \frac{s_2^2}{n_2}}$$

3. Finally, a 95% confidence interval for the difference of population means, $\mu_1 - \mu_2$, can be calculated from the following formula

$$(\bar{x}_1 - \bar{x}_2) \pm (\text{coefficient})SE(\bar{x}_1 - \bar{x}_2)$$

where the coefficient is 1.96 if $n_1 + n_2$ is large; otherwise, a t-coefficient is used with approximately

$$df = n_1 + n_2 - 2$$

Example 4.6

In Example 4.4 we have the following figures for a sample of joggers:

$$n_1 = 25$$
$$\bar{x}_1 = 47.5 \text{ ml/kg}$$
$$s_1^2 = 23.04$$

In addition, we have these data from a second sample consisting of 26 non-joggers:

$$n_2 = 26$$
$$\bar{x}_2 = 37.5$$
$$s_2^2 = 26.01$$

To proceed with the computation of a confidence interval for the difference of two population means, we have

$$\text{SE}(\bar{x}_1 - \bar{x}_2) = \sqrt{\frac{23.04}{25} + \frac{26.01}{26}}$$
$$= 1.39$$

It follows that a 95% confidence interval for the difference of means is

$$(47.5 - 37.5) \pm (1.96)(1.39) = (7.28, 12.72) \text{ ml/kg}$$

Since both limits of the confidence interval are positive, the interval does not include the value zero; this means that joggers are almost sure (we are 95% confident) to have a higher average VO_2 uptake than non-joggers.

An alternative, but less often used, procedure is to use as standard error of the difference of means

$$\sqrt{s_p^2 \left(\frac{1}{n_1} + \frac{1}{n_2} \right)}$$

where s_p^2 is the weighted average of s_1^2 and s_2^2,

$$s_p^2 = \frac{(n_1 - 1)s_1^2 + (n_2 - 1)s_2^2}{(n_1 - 1) + (n_2 - 1)}$$

The two procedures produce similar results most of the time.

4.3. ESTIMATION OF PROPORTIONS AND ODDS RATIOS

The sample proportion

$$p = \frac{x}{n}$$

where x is the number of positive outcomes and n is the sample size, can also be expressed as

$$p = \frac{\sum x_i}{n}$$

where x_i is "1" if the ith outcome is positive and "0" otherwise. In other words, a sample proportion can be viewed as a special case of sample means where data are coded as 0/1; and because of this, the Central Limit Theorem applies: The sampling distribution of p will be approximately normal when the sample size n is large. The mean and variance of this sampling distribution are

$$\mu_p = \pi$$

and

$$\sigma_p^2 = \frac{\pi(1-\pi)}{n}$$

respectively, where π is the population proportion.

Example 4.7

Suppose the true proportion of smokers in a community is known to be in the vicinity of $\pi = .4$, and we want to estimate it using a sample of size $n = 100$. The Central Limit Theorem indicates that p follows a normal distribution with mean

$$\mu_p = .40$$

and variance

$$\sigma_p^2 = \frac{(.4)(.6)}{100}$$

or standard deviation

$$\sigma_p = .049$$

Suppose we want our estimate to be correct within $\pm 3\%$; it follows that

$$\Pr(.37 \le p \le .43) = \Pr\left(\frac{.37 - .40}{.049} \le z \le \frac{.43 - .40}{.049}\right)$$

$$= \Pr(-.61 \le z \le .61)$$

$$= (2)(.2291)$$

$$= .4582, \text{ or approximately } 46\%$$

That means if we use the proportion of smokers from a sample of $n = 100$ to estimate the true proportion of smokers, only about 46% of the time we are correct within $\pm 3\%$; this figure would be 95.5% if the sample size is raised to $n = 1,000$. What we learn from this example is that, compared with the case of continuous data in Example 4.2, it may take a much larger sample to have a good estimate of a proportion such as a disease prevalence or a drug side effect.

4.3.1. Confidence Interval for a Proportion

Similar to the case of the mean, an approximate 95% confidence interval for a population proportion π is given by

$$p \pm (1.96)\text{SE}(p)$$

where the standard error of the sample proportion is calculated from

$$\text{SE}(p) = \sqrt{p(1-p)/n}$$

There are no easy ways for small samples; this is applicable only to larger samples ($n > 25$, n should be much larger for a narrow interval; procedures for small samples are rather complicated and are not covered in this book).

Example 4.8

Consider the problem of estimating the prevalence of malignant melanoma in 45–54-year-old women in the United States. Suppose a random sample of $n = 5,000$ women is selected from this age group and $x = 28$ are found to have the disease. Our point estimate for the prevalence of this disease is

$$p = \frac{28}{5,000}$$

$$= .0056$$

Its standard error is

$$SE(p) = \sqrt{(.0056)(1 - .0056)/5,000}$$
$$= .0011$$

Therefore, a 95% confidence interval for the prevalence π of malignant melanoma in 45–54-year-old women in the United States is given by

$$.0056 \pm (1.96)(.0011) = (.0034, .0078)$$

4.3.2. Difference of Proportions

For the comparisons of proportions using two independent samples, we have a similar formula for a 95% confidence interval for the difference $(\pi_1 - \pi_2)$,

$$(p_1 - p_2) \pm (1.96)SE(p_1 - p_2)$$

Since the variance of a sample proportion is

$$\sigma_p^2 = \frac{\pi(1 - \pi)}{n}$$

the standard error of the difference of sample proportions is given by

$$SE(p_1 - p_2) = \sqrt{p_1(1 - p_1)/n_1 + p_2(1 - p_2)/n_2}$$

Example 4.9

A public health official wishes to know how effective health education efforts are regarding smoking. Of $n_1 = 100$ males sampled in 1965 at the time of the release of the Surgeon General's Report on the health consequences of smoking, $x_1 = 51$ were found to be smokers. In 1980 a second random sample of $n_2 = 100$ males, similarly gathered, indicated that $x_2 = 43$ were smokers. Application of the above method yields

$$p_1 = \frac{51}{100}$$
$$= .51$$

$$p_2 = \frac{43}{100}$$

$$= .43$$

$$p_1 - p_2 = .080$$

$$SE(p_1 - p_2) = \sqrt{\frac{(.51)(.49)}{100} + \frac{(.43)(.57)}{100}}$$

$$= .070$$

Therefore, a 95% confidence interval for the change in smoking prevalence is

$$.080 \pm (1.96)(.070) = .080 \pm 1.38$$

$$= (-.058, .218)$$

These figures would give the public health official 95% confidence that the change in percentage of male smokers is between an increase of 5.8% to a reduction of 21.8% over the 15-year period. The reason for the wide interval is the small sample sizes. If we observe the same proportions ($p_1 = .51$ and $p_2 = .43$) using sample sizes $n_1 = n_2 = 1,000$, the interval would be (.036, .124) showing a decrease of between 3.6% and 12.4%.

_____

_____

Example 4.10

A study was conducted to look at the effects of oral contraceptives (OC) on heart disease in women 40–44 years of age. It is found that among $n_1 = 5,000$ current OC users, 13 develop a myocardial infarction (MI) over a 3-year period, while among $n_2 = 10,000$ non-OC users, 7 develop an MI over a 3-year period. Application of the above method yields

$$p_1 = \frac{13}{5,000}$$

$$= .0026$$

$$p_2 = \frac{7}{10,000}$$

$$= .0007$$

$$p_1 - p_2 = .0019$$

$$\text{SE}(p_1 - p_2) = \sqrt{\frac{(.0026)(.9974)}{5,000} + \frac{(.0007)(.9993)}{10,000}}$$

$$= .0008$$

Therefore, a 95% confidence interval for the difference of true proportions is

$$.0019 \pm (1.96)(.0008) = (.0004, .0034)$$

showing a small but positive difference. (The difference, .0019, is small because the risks themselves are low, $p_1 = .0026$ and $p_2 = .0007$; even the upper limit of the interval, i.e., .0034, is very small.)

4.3.3. Methods for Odds Ratios

There are different methods for the calculation of confidence intervals for odds ratios; most of these produce similar results, at least when the point estimate is not too extreme. We will present only one of these methods here.

Data from a case–control study, for example, may be summarized in a 2×2 table as shown in Table 4.3. The observed odds ratio (OR) is

$$\text{OR} = \frac{ad}{bc}$$

Confidence intervals are derived from the normal approximation to the sampling distribution of ln(OR) with

$$\text{Variance}[\ln(\text{OR})] \cong \frac{1}{a} + \frac{1}{b} + \frac{1}{c} + \frac{1}{d}$$

Consequently, an approximate 95% confidence interval for odds ratio is given by *exponentiating* (the reverse log operation) the two numbers

$$\ln\frac{ad}{bc} \pm 1.96\sqrt{\frac{1}{a} + \frac{1}{b} + \frac{1}{c} + \frac{1}{d}}$$

(ln is logarithm to base e, also called the *natural* logarithm.)

TABLE 4.3 2×2 Table Showing Data From a Case–Control Study

	Exposed	Unexposed
Diseased	a	b
Disease-free	c	d

Example 4.11

The role of smoking in pancreatitis has been recognized for many years; the following are data from a case–control study carried out in Eastern Massachusetts and Rhode Island (1975–79; see Example 1.10).

Use of cigarettes	Cases	Controls
Current smokers	38	81
Ex-smokers	13	80
Never	2	56

We have

(i) For ex-smokers, compared with those who have never smoked,

$$OR = \frac{(13)(56)}{(80)(2)}$$

$$= 4.55$$

and a 95% confidence interval for the population odds ratio is

$$\mathrm{Exp}\left[\ln 4.55 \pm 1.96\sqrt{\frac{1}{13} + \frac{1}{56} + \frac{1}{80} + \frac{1}{2}}\right] = (.99, 20.96)$$

(ii) For current smokers, compared with those who have never smoked,

$$OR = \frac{(38)(56)}{(81)(2)}$$

$$= 13.14$$

and a 95% confidence interval is

$$\mathrm{Exp}\left[\ln 13.14 \pm 1.96\sqrt{\frac{1}{38} + \frac{1}{56} + \frac{1}{81} + \frac{1}{2}}\right] = (3.04, 56.70)$$

(Here, "Exp" is the exponentiation, or anti-natural log, operation.)

EXERCISES

1. Consider a population consisting of four subjects: A, B, C, and D, with the following values for a random variable X under investigation:

Subject	Value
A	1
B	1
C	0
D	0

Form the sampling distribution for the sample mean of size $n = 2$ and verify that

$$\mu_{\bar{x}} = \mu$$

(sampling without replacement).

2. The body mass index (kg/m^2) is calculated by dividing a person's weight by the square of his or her height and is used as a measure of the extent to which the individual is overweight. Suppose the distribution of the body mass index for men has a standard deviation of $\sigma = 3$ kg/m^2, and we wish to estimate the mean μ using a sample of size $n = 49$. Find the probability that we would be correct within 1 kg/m^2.

3. The data on percentage saturation of bile for 29 female patients are given in Exercise 8 of Chapter 2. Find a 95% confidence interval for the population mean (try both with 1.96 and with the t-coefficient).

4. Using the weights of 57 children (Example 2.2), calculate a 95% confidence interval for the mean.

5. In a study of water pollution, a sample of mussels was taken and lead concentrations (milligrams per gram dry weight) was measured for each mussel. The following data were obtained:

$$\{113.0, 140.5, 163.3, 185.7, 202.5, 207.2\}$$

Calculate the 95% confidence interval for the mean μ.

6. Consider the data taken from a study that examines the response to ozone and sulfur dioxide among adolescents suffering from asthma. The following are measurements of forced expiratory volume (liters) for 10 subjects:

$$\{3.50, 2.60, 2.75, 2.82, 4.05, 2.25, 2.68, 3.00, 4.02, 2.85\}$$

Calculate the 95% confidence interval for the mean.

7. The percentage of ideal body weight was determined for 18 randomly selected insulin-dependent diabetics. The outcomes (%) are

107	119	99	114	120	104	124	88	114
116	101	121	152	125	100	114	95	117

Calculate the 95% confidence interval for the mean.

8. From the data on percentage saturation of bile in men (used in Example 4.3) and women (used in Exercise 6, Chapter 2; see also Exercise 4.2), calculate a 95% confidence interval for the difference of means (male–female).

9. From the data of the screening test for glaucoma in Exercise 6, Chapter 1, find a 95% confidence interval for the true specificity (specificity is a special use of proportion).

10. From the data of the case–control study for breast cancer among women over 54 years of age in Exercise 16 of Chapter 1, find a 95% confidence interval for

   (a) The proportion among the controls who consumed at least 150,500 international units of vitamin A per month

   (b) the odds ratio

11. Using the data in Example 4.10, calculate a 95% confidence interval for the odds ratio.

12. Using the data in Example 4.11, calculate a 95% confidence interval for the difference (case–control) of the proportions of non-smokers.

13. A study was conducted to investigate drinking problems among college students. In 1983, a group of students were asked whether they had ever driven an automobile while drinking. In 1987, after the legal drinking age was raised, a different group of college students were asked the same question. The results are as follows:

Drove while drinking	Year		Total
	1983	1987	
Yes	1,250	991	2,241
No	1,387	1,666	3,053
Total	2,637	2,657	5,294

Calculate the 95% confidence interval for the difference of two proportions: (1987)–(1983).

14. A study was undertaken to clarify the relationship between heart disease and occupational carbon disulfide exposure along with another important factor, elevated diastolic blood pressure (DBP), in a data set obtained from a 10-year prospective follow-up of two cohorts of over 340 male industrial workers in Finland. Carbon disulfide is an industrial solvent that is used all over the world in the production of viscose rayon fibers. The following table gives the mean and standard deviation (SD) of serum cholesterol (mg/100 ml) among exposed and nonexposed cohorts, by diastolic blood pressure (DBP).

DBP (mmHg)	Exposed			Nonexposed		
	n	Mean	SD	n	Mean	SD
< 95	205	220	50	271	221	42
95–100	92	227	57	53	236	46
≥ 100	20	233	41	10	216	48

Compare serum cholesterol levels between exposed and nonexposed cohorts at each level of DBP by calculating 95% confidence intervals for the difference of means (exposed–nonexposed). Any difference in these DBP-specific differences would indicate an effect modification.

15. A study was undertaken to investigate the roles of blood-borne environmental exposures on ovarian cancer from assessment of consumption of coffee, tobacco, and alcohol. Study subjects consist of 188 women in the San Francisco Bay area with epithelial ovarian cancers diagnosed in 1983–1985 and 539 control women. Of the 539 controls, 280 were hospitalized women without overt cancer, and 259 were chosen from the general population by random telephone dialing. Data for coffee consumption are summarized as follows:

Coffee drinkers	Cases	Hospital controls	Population controls
No	11	31	26
Yes	177	249	233

Calculate the odds ratio and its 95% confidence interval for

(a) Cases versus hospital controls

(b) Cases versus population controls

16. The prevalence rates of hypertension among adult (ages 18–74) white and black Americans were measured in the second National Health and Nutrition Exam-

ination Survey, 1976–80. Prevalence estimates (and their standard errors) for women are given below:

	%	SE
Whites	25.3	0.9
Blacks	38.6	1.8

Calculate and compare the 95% confidence intervals for the proportions of the two groups and draw a conclusion.

17. An experimental study was conducted with 136 5-year-old children in four Quebec schools to investigate the impact of simulation games designed to teach children to obey certain traffic safety rules. The transfer of learning was measured by observing children's reactions to a quasi-real-life model of traffic risks. The scores on the transfer of learning for the control and attitude/behavior simulation game groups are summarized below:

Summarized data	Control	Simulation game
n	30	33
$\bar{x}$	7.9	10.1
s	3.7	2.3

Find and compare the 95% confidence intervals for the means of the two groups and draw a conclusion.

18. The body mass index is calculated by dividing a person's weight by the square of his or her height (it is used as a measure of the extent to which the individual is overweight). A sample of 58 men, selected (retrospectively) from a large group of middle-aged men who later developed diabetes mellitus, yields $\bar{x} = 25.0$ kg/m^2 and $s = 2.7$ kg/m^2.

(a) Calculate a 95% confidence interval for the mean of this sub-population.

(b) If it is known that the average body mass index for middle-aged men who do not develop diabetes is 24.0 kg/m^2, what can you say about the relationship between body mass index and diabetes in middle-aged men?

19. Postneonatal mortality due to respiratory illnesses is known to be inversely related to maternal age, but the role of young motherhood as a risk factor for respiratory morbidity in infants has not been thoroughly explored. A study was conducted in Tucson, Arizona, aimed at the incidence of lower respiratory tract illnesses during the first year of life. In this study, over 1,200 infants were en-

rolled at birth between 1980 and 1984, and the following data are concerned with wheezing lower respiratory tract illnesses (wheezing LRI): No/Yes.

Maternal age (years)	Boys		Girls	
	No	Yes	No	Yes
< 21	19	8	20	7
21–25	98	40	128	36
26–30	160	45	148	42
> 30	110	20	116	25

Using "> 30" as the baseline, calculate the odds ratio and its 95% confidence interval for each other maternal age group.

20. Data were collected from 2,197 white ovarian cancer patients and 8,893 white controls in 12 different U.S. case–control studies conducted by various investigators in the period 1956–1986. These were used to evaluate the relationship of invasive epithelial ovarian cancer to reproductive and menstrual characteristics, exogenous estrogen use, and prior pelvic surgeries. The following are parts of the data:

(a)

Duration of unprotected intercourse (years)	Cases	Controls
< 2	237	477
2–9	166	354
10–14	47	91
≥ 15	133	174

Using "< 2" as the baseline, calculate the odds ratio and its 95% confidence interval for each other level of exposure.

(b)

History of infertility	Cases	Controls
No	526	966
Yes		
No drug use	76	124
Drug use	20	11

Using "no history of infertility" as the baseline, calculate the odds ratio and its 95% confidence interval for each group with a history of infertility.

CHAPTER

5

Introduction to Hypothesis Testing

This chapter covers the most used and yet most misunderstood statistical procedures, called *tests* or *tests of significance*. The less people understand about statistics the more they think statistical tests are good things to do to data. The reason for the misunderstanding is simple: language. The colloquial meaning of the word *test* is one of no-nonsense objectivity. Students take tests in school, hospitals draw blood to be sent to laboratories for tests, and automobiles are tested by the manufacturer for performance and safety. It is thus natural to think that statistical tests are the "objective" procedures to use on data. It just happens to be false. Statisticians have created the misunderstanding by using the word *test* for this one class of statistical procedures. Statistical tests are no more or less objective than any other statistical procedure.

Statisticians have made the problem worse by using the word *significance*. Significance is another word that has a powerful meaning in ordinary, colloquial language: *importance*. Statistical tests that result in "significance" are naturally misunderstood by the public to mean that the data are statistically important. That's not what statisticians mean. Every competent statistician knows that statistical significance does not mean statistical importance. If someone tells you they think "statistical significance" means "statistical importance" you know that person is not a competent statistician.

Statistical tests are commonly and seriously misinterpreted by non-statisticians, but the misinterpretations are very natural. It is very natural to look at data and ask whether there is "anything going on" or whether it is just a bunch of meaningless numbers that can't be interpreted. Statistical tests appeal to the non-statistician for a reason in addition to the aforementioned reasons of language confusion. Statistical tests are appealing because they seem to make a decision; they are attractive because they say "yes" or "no." There is comfort in using a procedure that gives definitive answers from confusing data.

One way of explaining statistical tests, used by many teachers of statistics who teach courses at this level, is to use criminal court procedures as a metaphor. In criminal court, the accused is "presumed innocent" until "proven guilty beyond all reasonable doubt." This framework of presumed innocence has nothing whatsoever to do with anyone's personal belief as to the innocence or guilt of the defendant. Sometimes everybody in their right mind, including the jury, the judge, and even the defendant's attorney think the defendant is guilty as sin. The rules and procedures of criminal court, however, must be followed. There may be a mistrial, or a hung jury, or the arresting officer forgot to read the defendant his or her rights. Any number of things can happen to save the guilty from a conviction. On the other hand, an innocent defendant is sometimes convicted by overwhelming circumstantial evidence. Criminal courts occasionally make mistakes, sometimes releasing the guilty and sometimes convicting the innocent. Statistical tests are like that. Sometimes statistical significance is attained when nothing is going on and sometimes no statistical significance is attained when something very important is going on.

Just as in the court room, everyone would like statistical tests to make mistakes as infrequently as possible. Actually, the mistake rate of one of the kinds of mistakes made by statistical tests has been (arbitrarily) chosen to be 5%. The kind of mistake referred to here is the mistake of attaining statistical significance when there is actually nothing going on. This mistake is called a Type I mistake. Statistical tests are constructed so that Type I mistakes occur 5% of the time. There is no custom regarding the rate of Type II mistakes, however. A Type II mistake is the mistake of not getting statistical significance when there is something going on. The rate of Type II mistakes is dependent on several factors. One of the factors is *how much* is going on. If there is a lot going on, one is less likely to make Type II mistakes. Another factor is the amount of variability ("noise") there is in the data. A lot of variability makes Type II mistakes more likely. Yet another factor is the size of the study. There are more Type II mistakes in small studies than there are in large ones. Type II mistakes are rare in really huge studies, but quite common in small studies.

There is a very important, subtle aspect of statistical tests, based on the aforementioned three things that make Type II mistakes very improbable. Since really huge studies virtually guarantee getting statistical significance if there is even the slightest amount going on, such studies result in statistical significance when the *amount* that is going on is of no practical importance. In this case, statistical significance is attained in the face of no practical significance. On the other hand, small studies can result in statistical *non*-significance when something of great practical importance is going on. The conclusion is that the attainment of statistical significance in a study is just as affected by extraneous factors as it is by practical importance. It is essential to learn that *statistical significance* is not synonymous with *practical importance*. Every statistician knows this fact.

Granting all these subtle aspects of statistical tests, and knowing that it is easy to learn how to do them, we have to include them in this text. Most non-statisticians, and even many statisticians, use them all the time. Since they are so commonly done, and even *required* by some government regulatory agencies, we would be remiss as text writers and teachers if we did not include them. We want you to learn how to do some of the more common, elementary ones typically included in courses at this level. We also, however, urge that you interpret them with caution and seek expert help when you are unsure of their interpretation.

5.1. BASIC CONCEPTS

From the introduction of sampling distributions in Chapter 4, it was clear that the value of a sample mean is influenced by

(i) The population μ, because

$$\mu_{\bar{x}} = \mu$$

(ii) Chance; $\bar{x}$ and μ are almost never identical. The variance of the sampling distribution is

$$\sigma_{\bar{x}}^2 = \frac{\sigma^2}{n}$$

a combined effect of natural variation in the population (σ^2) and sample size (n).

Therefore, when an observed value $\bar{x}$ is far from a hypothesized value of μ (e.g., mean high blood pressures for a group of oral contraceptive users compared with a typical average), a natural question would be "Was it just due to chance, or something else?" To deal with questions such as this, statisticians have invented the concept of *hypothesis tests*, and these tests have become widely used statistical techniques in the health sciences. In fact, it is almost impossible to read a research article in public health or medical sciences without running across hypothesis tests!

5.1.1. Hypothesis Tests

When a health investigator seeks to understand or explain something, for example, the effect of a toxin or a drug, he or she usually formulates his or her research question in the form of a *hypothesis*. In the statistical context, a hypothesis is a statement about a distribution (e.g., "the distribution is normal") or its underlying parameter(s) (e.g., "$\mu = 10$"), or about the relationship between probability distributions (e.g., "there is no statistical relationship") or its parameters (e.g., "$\mu_1 = \mu_2$"— equality of population means). The hypothesis to be tested is called the *null hypothesis* and will be denoted by $\mathcal{H}_0$; it is usually stated in the "null" form, indicating no difference or no relationship between distributions or parameters. In other words, under the null hypothesis, an observed difference (like the one between sample means $\bar{x}_1$ and $\bar{x}_2$ for Sample 1 and Sample 2, respectively) just reflects chance variation. A *hypothesis* test is a decision-making process that examines a set or sets of data and, on the basis of expectation under $\mathcal{H}_0$, leads to a decision on whether or not to reject $\mathcal{H}_0$.

An *alternative hypothesis*, which we denote by $\mathcal{H}_A$, is a hypothesis that in some sense contradicts the null hypothesis $\mathcal{H}_0$. Under $\mathcal{H}_A$, the observed difference is real (e.g., $\bar{x}_1 \neq \bar{x}_2$ not by chance but because $\mu_1 \neq \mu_2$). A null hypothesis is rejected if and only if there is sufficiently strong evidence from the data to support its alternative—the names are somewhat unsettling, because the "alternative hypothesis" is, for a health investigator, the one he or she usually wants to prove (the null hypothesis is just a dull explanation of the findings—in terms of chance variation!). However, these are entrenched statistical terms and will be used as standard terms for the rest of this book.

Why is hypothesis testing important? Because in many circumstances we merely wish to know whether a certain proposition is true or false. The process of hypothesis tests provides a framework for making decisions on an *objective* basis, by weighing the relative merits of different hypotheses, rather than a *subjective* basis by simply looking at the numbers. Different people can form different opinions by looking at data (confounded by chance variation or sampling errors), but a hypothesis test provides a standardized decision-making process that will be consistent for all people. The mechanics of the tests vary with the hypotheses and measurement scales (see Chapter 6 and parts of Chapter 7), but the general philosophy and foundation is common and will be discussed with some detail in this chapter.

5.1.2. Statistical Evidence

A null hypothesis is often concerned with parameter(s) of population(s). However, it is often either impossible or too costly or time-consuming to obtain the entire population data on any variable in order to see whether or not a null hypothesis is true. Decisions are thus made using sample data. Sample data are summarized into a statistic or statistics that are then used to estimate the parameter(s) involved in the null hypothesis. For example, if a null hypothesis is about μ (e.g., $\mathcal{H}_0 : \mu = 10$), then a good place to look for an estimate of μ is $\bar{x}$. In that context, the statistic $\bar{x}$ is called a *test statistic*; a test statistic measures the difference between the data (i.e., the numerical value of $\bar{x}$ obtained from the sample) and what is expected if the null hypothesis is true (i.e., "$\mu = 10$"). However, this evidence is statistical evidence; it varies from sample to sample (in the context of repeated sampling). It is a variable with a specific sampling distribution. The observed value is thus usually converted to a standard unit: the number of standard errors away from a hypothesized value. At this point, the logic of the test can be seen more clearly. It is an argument by contradiction, designed to show that the null hypothesis will lead to a less acceptable conclusion (an almost impossible event—some event that occurs with near zero probability) and must therefore be rejected. In other words,

TABLE 5.1   Possible Outcomes of a Statistical Test

	Decision	
Truth	$\mathcal{H}_0$ **is not rejected**	$\mathcal{H}_0$ **is rejected**
$\mathcal{H}_0$ is true	Correct decision	Type I error
$\mathcal{H}_0$ is false	Type II error	Correct decision

the difference between the data and what is expected on the null hypothesis would be very difficult—even absurd—to explain as a chance variation; it makes you want to abandon (or reject) the null hypothesis and believe in the alternative hypothesis.

5.1.3. Errors

Since a null hypothesis $\mathcal{H}_0$ may be true or false and our possible decisions are whether to reject or not to reject it, there are four possible outcomes or combinations. Two of the four outcomes are correct decisions:

(i) Not rejecting a true $\mathcal{H}_0$
(ii) Rejecting a false $\mathcal{H}_0$

but there are also two possible ways to commit an error:

(i) *Type I*: a true $\mathcal{H}_0$ is rejected
(ii) *Type II*: a false $\mathcal{H}_0$ is not rejected

These possibilities are shown in Table 5.1. The general aim in hypothesis testing is to keep α and β, the probabilities—in the context of repeated sampling—of types I and II, respectively, as small as possible. However, if resources are limited, this goal requires a compromise because these actions are contradictory; conventionally, we fix α at some specific conventional level—say .05 or .01—and β is controlled through the use of sample size(s).

Example 5.1

Suppose the national smoking rate among men is 25% and we want to study the smoking rate among men in the New England states. Let π be the proportion of New England men who smoke. The null hypothesis that the smoking prevalence in New England is the same as the national rate is expressed as

$$\mathcal{H}_0 : \pi = .25$$

Suppose we plan to take a sample of size $n = 100$ and use this decision-making rule:

"If $p \leq .20$, then $\mathcal{H}_0$ is rejected"

where p is the proportion obtained from the sample.

(i) Alpha (α) is defined as the probability of wrongly rejecting a true null hypothesis, that is,

$$\alpha = \Pr(p \leq .20, \text{ given that } \pi = .25)$$

Since $n = 100$ is large enough for the Central Limit Theorem to apply, the sampling distribution of p is approximately normal with mean and variance, under $\mathcal{H}_0$, given by

$$\mu_p = \pi$$

$$= .25$$

$$\sigma_p^2 = \frac{\pi(1-\pi)}{n}$$

$$= (.043)^2$$

respectively. Therefore, for this decision-making rule,

$$\alpha = \Pr\left(z \leq \frac{.20 - .25}{.043}\right)$$

$$= \Pr(z \leq -1.16)$$

$$= .123 \text{ or } 12.3\%$$

Of course, we can make this smaller (as small as we wish) by changing the decision-making rule; however, that action will increase the value of β (or the probability of a Type II error).

(ii) Suppose that the truth is

$$\mathcal{H}_A : \pi = .15$$

Beta (β) is defined as the probability of not rejecting a false $\mathcal{H}_0$, that is,

$$\beta = \Pr(p > .20; \text{ knowing that } \pi = .15)$$

Again, an application of the Central Limit Theorem indicates that the sampling distribution of p is approximately normal with mean

$$\mu_p = .15$$

and variance

$$\sigma_p^2 = \frac{(.15)(.85)}{100}$$
$$= (.036)^2$$

Therefore

$$\beta = \Pr\left(z \geq \frac{.20 - .15}{.036}\right)$$
$$= \Pr(z \geq 1.39)$$
$$= .082 \text{ or } 8.2\%$$

The above results can be represented graphically as follows:

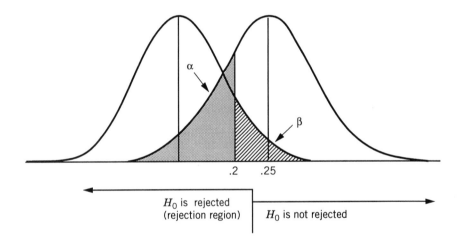

It can be seen that β depends on a specific alternative (e.g., β is larger for $\mathcal{H}_A$: $\pi = .17$ or any alternative hypothesis that specifies a value of π that is nearer to .25), and, from the above graph, if we change the decision making rule by using a smaller "cut point," then we would decrease α but increase β.

5.2. ANALOGIES

To reinforce some of the definitions or terms we have encountered, we consider in this section two analogies: trials by jury and medical screening tests.

5.2.1. Trials by Jury

Statisticians and statistics users may find a lot in common between a court trial and a statistical test of significance. In a criminal court, the jury's duty is to evaluate the evidence of the prosecution and the defense to determine whether a defendant is guilty or innocent. By use of the judge's instructions, which provide guidelines for their reaching a decision, the members of the jury can arrive at one of two verdicts: guilty or not guilty. Their decision may be correct or they could make one of two possible errors: convict an innocent person or free a criminal. The analogy between statistics and trials by jury goes as follows:

Test of significance ⇔ Court trial

Null hypothesis ⇔ "Every defendant is innocent

until proven guilty"

Research design ⇔ Police investigation

Data/test statistics ⇔ Evidence/exhibits

Statistical principles ⇔ Judge's instruction

Statistical decision ⇔ Verdict

Type I error ⇔ Conviction of an innocent defendant

Type II error ⇔ Acquittal of a criminal

This analogy clarifies a very important concept: When a null hypothesis is not rejected it does not necessarily lead to its acceptance, because a "not guilty" verdict is just an indication of "lack of evidence" and "innocence" is a possibility. That is, when a difference is not statistically significant, there are still two possibilities:

(i) The null hypothesis is true.
(ii) The null hypothesis is false, but there is not enough evidence from sample data to support its rejection (i.e., sample size is too small).

5.2.2. Medical Screening Tests

Another analogy of hypothesis testing can be found in the application of screening tests or diagnostic procedures. Following these procedures, clinical observations or laboratory techniques, individuals are classified as healthy or as having a disease. Of course, these tests are imperfect: Healthy individuals will occasionally be classified wrongly as being ill, while some individuals who are ill, may fail to be detected. The analogy between statistical tests and screening tests goes briefly as follows:

Type I error ⇔ False positives

Type II error ⇔ False negatives

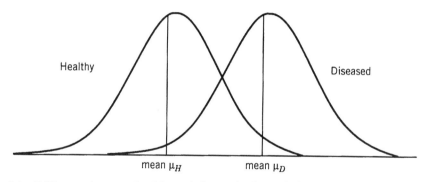

Healthy

Diseased

mean μ_H mean μ_D

Fig. 5.1 Difference between healthy and diseased subpopulations.

so that

$$\alpha = 1 - \text{specificity}$$

$$\beta = 1 - \text{sensitivity}$$

5.2.3. Common Expectations

The medical care system, with its high visibility and remarkable history of achievements, has been perceived somewhat naively by the general public as a perfect remedy factory. Medical tests are expected to diagnose correctly any disease (and physicians are expected to treat effectively and cure all diseases!). Another common misconception is the assumption that all tests, regardless of the disease being tested for, are equally accurate. People are shocked to learn that a test result is wrong (of course, the psychological effects could be devastating). Another analogy between tests of significance and screening tests exists here: Statistical tests are also expected to provide a correct decision!

In some medical cases such as infections, the presence or absence of bacteria and viruses are easier to confirm correctly. In other cases, such as the diagnosis of diabetes by a blood sugar test, the story is different. One very simple model for these situations would be to assume that the variable X (e.g., sugar level in blood) on which the test is based is distributed with different means for the healthy and diseased subpopulations (Figure 5.1).

It can be seen from Figure 5.1 that errors are unavoidable, especially when the two means μ_H and μ_D are close. The same is true for statistical tests of significance; when the null hypothesis $\mathcal{H}_0$ is not true, it could be wrong a little or it could be very wrong. For example, for

$$\mathcal{H}_0 : \mu = 10$$

the truth could be "$\mu = 12$" or "$\mu = 50$." If $\mu = 50$, Type II errors would be less likely, and if $\mu = 12$, Type II errors are more likely.

5.3. SUMMARIES AND CONCLUSIONS

To perform an hypothesis test we take the following steps:

1. Formulate a null hypothesis. (This would follow our research question, providing an explanation of what we want to prove in terms of chance variation.)
2. Design the experiment and obtain data.
3. Choose a test statistic. (This choice depends on the null hypothesis as well as measurement scale.)
4. Summarize findings and state appropriate conclusions.

This section involves the last step of the above process.

5.3.1. Rejection Region

The most common approach is the formation of a "Decision Rule." All possible values of the chosen test statistic (in the repeated sampling context) are divided into two regions. The region consisting of values of the test statistic for which the null hypothesis $\mathcal{H}_0$ is rejected is called the *rejection region*. The values of the test statistic comprising the rejection region are those values that are less likely to occur if the null hypothesis is true, and the Decision Rule tells us to reject $\mathcal{H}_0$ if the value of the test statistic that we compute from our sample(s) is one of the values in this region. For example, if a null hypothesis is about μ, say

$$\mathcal{H}_0 : \mu = 10$$

then a good place to look for a test statistic for $\mathcal{H}_0$ is $\bar{x}$, and it is obvious that $\mathcal{H}_0$ should be rejected if $\bar{x}$ is far away from "10," the hypothesized value of μ. Before we proceed, a number of related concepts should be made clear:

One-Tailed Versus Two-Tailed Tests

In the above example, a vital question is "Are we interested in the deviation of $\bar{x}$ from 10 in one or both directions?" If we are interested in determining whether μ is significantly *different* from 10, we would perform a two-tailed test and the rejection region would be as shown in Figure 5.2. On the other hand, if we are interested in

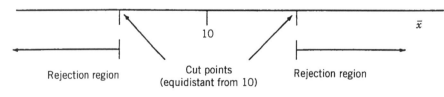

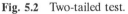

Fig. 5.2 Two-tailed test.

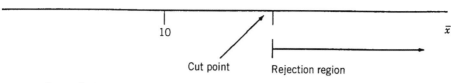

Fig. 5.3 One-tailed test.

whether μ is significantly *larger* than 10, we would perform a one-tailed test and the rejection region would be as shown in Figure 5.3.

A one-tailed test is indicated for research questions like these: Is a new drug *superior* to a standard drug? Does the air pollution *exceed* safe limits? Has the death rate been *reduced* for those who quit smoking? A two-tailed test is indicated for research questions like these: Is there a *difference* between cholesterol levels of men and women? Does the mean age of a target population *differ* from that of the general population?

Level of Significance

The decision as to which values of the test statistic go into the rejection region, or where the cut point is, is made on the basis of the desired level of Type I error α (also called the *size* of the test). A computed value of the test statistic that falls in the rejection region is said to be *statistically significant*. Common choices for α, the level of significance, are .01, .05, and .10.

Reproducibility

Here we aim to clarify another misconception about hypothesis tests. A very simple and common situation for hypothesis tests is that the test statistic, for example the sample mean $\bar{x}$, is normally distributed with different means under the null hypothesis $\mathcal{H}_0$ and alternative hypothesis $\mathcal{H}_A$. A one-tailed test could be graphically represented as in Figure 5.4.

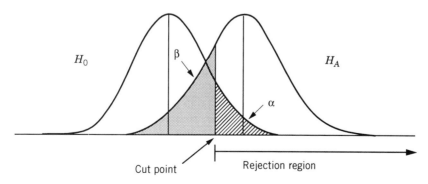

Fig. 5.4 Errors associated with a one-tailed test.

It should now be clear that a statistical conclusion is not guaranteed to be reproducible. For example if the alternative hypothesis is true and the mean of the distribution of the test statistic (see graph above) is right at the cut point, then the probability would be 50% to obtain a test statistic inside the rejection region.

5.3.2. *p*-Values

Instead of saying that an observed value of the test statistic is significant (i.e., falling into the rejection region for a given choice of α) or is not significant, many writers in the research literature prefer to report findings in terms of *p-values*. The *p*-value is the probability of getting values of the test statistic as extreme as, or more extreme than, that observed if the null hypothesis is true. For the above example of

$$\mathcal{H}_0 : \mu = 10$$

if the test is a one-tailed test, we would have the results shown in Figure 5.5. If the test is a two-tailed test then we have the results shown in Figure 5.6. The curve in Figures 5.5 and 5.6 represents the sampling distribution of $\overline{x}$ if $\mathcal{H}_0$ is true.

As compared to the approach of choosing a level of significance and formulating a decision rule, the use of the *p*-value criterion would be as follows:

(i) If $p < \alpha$, $\mathcal{H}_0$ is rejected
(ii) if $p \geq \alpha$, $\mathcal{H}_0$ is not rejected

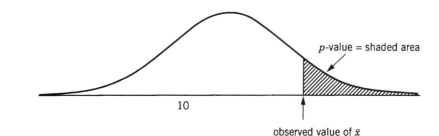

Fig. 5.5 *p*-value of a one-tailed test.

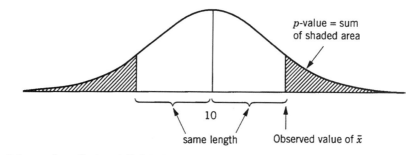

Fig. 5.6 *p*-value of a two-tailed test.

**TABLE 5.2 Conventional Interpretation of
p-Values**

p-value	Interpretation
$p > .10$	Result is not significant
$.05 < p < .10$	Result is marginally significant
$.01 < p < .05$	Result is significant
$p < .01$	Result is highly significant

However, the reporting of *p*-values as part of the results of an investigation is more informative to the readers than such statements as "the null hypothesis is rejected at the .05 level of significance" or "the results *were not significant at the .05 level." Reporting the *p*-value associated with a test lets the reader know how common or how rare is the computed value of the test statistic given that $\mathcal{H}_0$ is true. In other words, the *p*-value can be used as a *measure* of the compatibility between the data (reality) and a null hypothesis (theory); the smaller the *p*-value, the less compatible the theory and the reality. A compromise between the two approaches would be to report both in statements such as "the difference is statistically significant $(p < .05)$." In doing so, researchers generally agree on the conventional terms listed in Table 5.2. Finally, it should be noted that the difference between means, for example, although statistically significant may be so small that it has little health consequence. In other words, the result may be *statistically significant* but may **not** be *practically significant*.

Example 5.2

Suppose the national smoking rate among men is 25% and we want to study the smoking rate among men in the New England states. The null hypothesis under investigation is

$$\mathcal{H}_0 : \pi = .25$$

Of $n = 100$ males sampled, $x = 15$ were found to be smokers. Does the proportion π of smokers in New England states differ from that of the nation?

Since $n = 100$ is large enough for the Central Limit Theorem to apply, it indicates that the sampling distribution of the sample proportion p is approximately normal with mean and variance under $\mathcal{H}_0$:

$$\mu_p = .25$$
$$\sigma_p^2 = \frac{(.25)(1 - .25)}{100}$$
$$= (.043)^2$$

The observed value of p from our sample is

$$\frac{15}{100} = .15$$

representing a difference of .10 from the hypothesized value of .25. The p-value is defined as the probability of getting a value of the test statistic as extreme as, or more extreme than, that observed if the null hypothesis is true. This is represented graphically as follows:

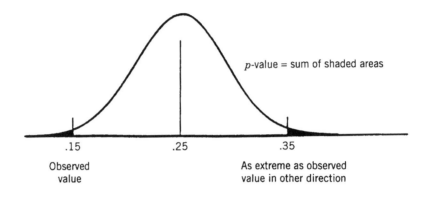

Therefore

$$p\text{-value} = \Pr(p \leq .15 \text{ or } p \geq .35)$$
$$= 2 \times \Pr(p \geq .35)$$
$$= 2 \times \Pr\left(z \geq \frac{.35 - .25}{.043}\right)$$
$$= 2 \times \Pr(z \geq 2.33)$$
$$= (2)(.5 - .4901)$$
$$\cong .02$$

In other words, with the data given, the difference between the national smoking rate and the smoking rate of New England states is statistically significant ($p < .05$).

5.3.3. Relationship to Confidence Intervals

Suppose we consider a hypothesis of the form

$$\mathcal{H}_0 : \mu = \mu_0$$

where μ_0 is a known hypothesized value. A two-tailed hypothesis test for $\mathcal{H}_0$ is related to confidence intervals as follows:

1. If μ_0 is not included in the 95% confidence interval for μ, $\mathcal{H}_0$ should be rejected at the .05 level. This is represented graphically in Figure 5.7.
2. If μ_0 is included in the 95% confidence interval for μ, $\mathcal{H}_0$ should not be rejected at the .05 level (Figure 5.8).

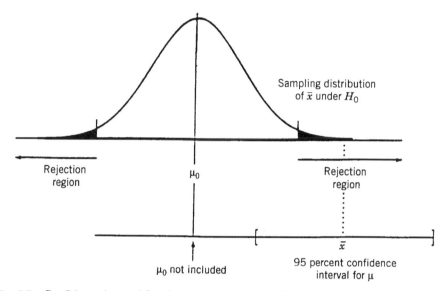

Fig. 5.7 Confidence interval for the mean where the null hypothesis is rejected.

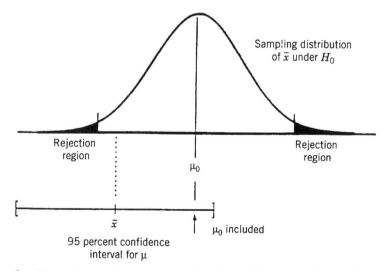

Fig. 5.8 Confidence interval for the mean where the null hypothesis is not rejected.

Example 5.3

Consider the hypothetical data set in Example 5.2. Our point estimate of smoking prevalence in New England is

$$p = \frac{15}{100}$$

$$= .15$$

Its standard error is

$$SE(p) = \sqrt{(.15)(1 - .15)/100}$$

$$= .036$$

Therefore, a 95% confidence interval for the New England states smoking rate π is given by

$$.15 \pm (1.96)(.036) = (.079, .221)$$

It is noted that the national rate of .25 is *not included* in that confidence interval.

EXERCISES

1. For each part, state the null ($\mathcal{H}_0$) and alternative ($\mathcal{H}_A$) hypotheses:
   (a) Has the average community level of suspended particulates for the month of August exceeded 30 mcg per cubic meter?
   (b) Does mean age of onset of a certain acute disease for school children differ from 11.5?
   (c) A psychologist claims that the average IQ of a sample of 60 children is significantly above the normal IQ of 100.
   (d) Is the average cross-sectional area of the lumen of coronary arteries for men, ages 40 to 59 years, less than 31.5% of the total arterial cross section?
   (e) Is the mean hemoglobin level of a group of high-altitude workers different from 16 g/cc?
   (f) Does the average speed of 50 cars as checked by radar on a particular highway differ from 55 mph?
2. The distribution of diastolic blood pressures for the population of female diabetics between the ages of 30 and 34 years has an unknown mean μ and a standard deviation of $\sigma = 9$ mmHg. It may be useful to physicians to know whether

the mean μ of this population is equal to the mean diastolic blood pressure of the general population of females of this age group, which is 74.5 mmHg. What is the null hypothesis, and what is the alternative hypothesis for this test?

3. *E. canis* infection is a tick-borne disease of dogs that is sometimes contracted by humans. Among infected humans, the distribution of white blood cell counts has an unknown mean μ and a standard deviation σ. In the general population the mean white blood count is $7,250/mm^3$. It is believed that persons infected with *E. canis* must on average have a lower white blood cell count. What is the null hypothesis for the test? Is this a one-sided or two-sided alternative?

4. It is feared that the smoking rate in young females has increased in the last several years. In 1985, 38% of the females in the 17–24-year age group were smokers. An experiment is to be conducted to gain evidence to support the increase contention. Set up the appropriate null and alternative hypotheses. Explain in a practical sense what, if anything, has occurred if a Type I or a Type II error has been committed.

5. A group of investigators wishes to explore the relationship between the use of hair dyes and the development of breast cancer in females. A group of 1,000 beauticians 40–49 years of age is identified and followed for 5 years. After 5 years, 20 new cases of breast cancer have occurred. Assume that breast cancer incidence over this time period for average American women in this age group is 7/1,000. We wish to test the hypothesis that using hair dyes increases the risk of breast cancer. Is a one-sided or a two-sided test appropriate here? Compute the *p*-value for your choice.

6. Height and weight are often used in epidemiologic studies as possible predictors of disease outcomes. If the people in the study are assessed in a clinic, then heights and weights are usually measured directly. However, if the people are interviewed at home or by mail, then a person's self-reported height and weight are often used instead. Suppose we conduct a study on 10 people to test the comparability of these two methods. Data from these 10 people were obtained using both methods on each person. What is the criterion for the comparison? What is the null hypothesis? Should a two-sided or a one-sided test be used here?

7. Suppose that 28 cancer deaths are noted among workers exposed to asbestos in a building materials plant from 1981 to 1985. Only 20.5 cancer deaths are expected from statewide mortality rates. Suppose we want to know if there is a significant excess of cancer deaths among these workers. What is the null hypothesis? Is a one-sided or two-sided test appropriate here?

8. A food frequency questionnaire was mailed to 20 subjects to assess the intake of various food groups. The sample standard deviation of vitamin C intake over the 20 subjects was 15 (exclusive of vitamin C supplements). Suppose we know from using an in-person diet interview method in a large previous study that the standard deviation is 20. Formulate the null and alternative hypotheses if we want to test for any differences between the standard deviations of the two methods. (Note: comparisons of standard deviations of variances are not covered in this book.)

9. In Example 5.1, it was assumed that the national smoking rate among men is 25%. A study is to be conducted for New England states using a sample size $n = 100$ and the decision rule

"If $p \leq .20$, $\mathcal{H}_0$ is rejected"

where $\mathcal{H}_0$ is

$$\mathcal{H}_0 : \pi = .25$$

where π and p are population and sample proportions, respectively, for New England states. Is this a one-tailed or two-tailed test?

10. In Example 5.1, with the rule

"If $p \leq .20$, $\mathcal{H}_0$ is rejected"

it was found that the probabilities of Type I and Type II errors are

$$\alpha = .123$$

$$\beta = .082$$

for $\mathcal{H}_A : \pi = .15$. Find α and β if the rule is changed to

"If $p \leq .18$, $\mathcal{H}_0$ is rejected"

How does this change affect α and β values?

11. Answer the questions in Exercise 5.10 above for the decision rule

"If $p \leq .22$, $\mathcal{H}_0$ is rejected"

12. Recalculate the p-value in Example 5.2 if it was found that 18 (instead of 15) men in a sample of $n = 100$ are smokers.

13. Calculate the 95% confidence interval for π using the sample in Exercise 5.12 above and compare the findings to the testing results of Exercise 5.9.

14. Plasma glucose levels are used to determine the presence of diabetes. Suppose the mean log plasma glucose concentration (mg/dl) in 35–44 year olds is 4.86 with standard deviation .54. A study of 100 sedentary persons in this age group is planned to test whether they have higher levels of plasma glucose than the general population.

(a) Set up the null and alternative hypotheses.

(b) If the real increase is .1 log units, then what is the power of such a study if a two-sided test is to be used with $\alpha = .05$?

15. Suppose we are interested in investigating the effect of race on level of blood pressure. The mean and standard deviation of systolic blood pressure among

25–34-year-old white males were reported as 128.6 mmHg and 11.1 mmHg, respectively, based on a very large sample. Suppose the actual mean for black males in the same age group is 135 mmHg. What is the power of the test (two-sided, $\alpha = .05$) if $n = 100$ and we assume that the variances for whites and blacks are the same?

6

Some Simple Statistical Tests

6.1. COMPARISONS OF PROPORTIONS

6.1.1. One-Sample Problem

In this type of problem, we have a sample of binary data (n, x), with n being an adequately large sample size and x the number of positive outcomes among the n observations, and we consider the null hypothesis

$$\mathcal{H}_0 : \pi = \pi_0$$

where π_0 is a fixed and known number between 0 and 1, for example,

$$\mathcal{H}_0 : \pi = .25$$

π_0 is often a standardized or referenced figure, for example, the effect of a standardized drug or therapy, or the national smoking rate (where the national sample is often large enough so as to produce negligible sampling error in π_0). Or, we could be concerned with a research question like "Does the side effect (of a certain drug) exceed regulated limit π_0?" In Exercise 5.2, we tried to compare the incidence of breast cancer among female beauticians (who are frequently exposed to the use of hair dyes) versus a standard level of 7/1,000 (for 5 years) for "average American women." The figure 7/1,000 is π_0 for that example.

To perform a test of significance for $\mathcal{H}_0$, we proceed with the following steps:

1. Decide whether a one-tailed or a two-tailed test is appropriate.
2. Choose a level of significance α, a common choice being .05.
3. Calculate the z-score

$$z = \frac{p - \pi_0}{\sqrt{\dfrac{\pi_0(1 - \pi_0)}{n}}}$$

4. From the table for the standard normal distribution (Appendix B) and the choice of α, e.g., $\alpha = .05$, the rejection region is determined by

- For a one-tailed test,

$$z \leq -1.65 \quad \text{for} \quad \mathcal{H}_A : \pi < \pi_0$$

$$z \geq 1.65 \quad \text{for} \quad \mathcal{H}_A : \pi > \pi_0$$

- For a two tailed test or $\mathcal{H}_A : \pi \neq \pi_0$,

$$z \leq -1.96 \quad \text{or} \quad z \geq 1.96$$

Example 6.1

A group of investigators wish to explore the relationship between the use of hair dyes and the development of breast cancer in women. A sample of $n = 1,000$ female beauticians 40–49 years of age is identified and followed for 5 years. After 5 years, $x = 20$ new cases of breast cancer have occurred. It is known that breast cancer incidence over this time period for average American women in this age group is $\pi_0 = 7/1,000$. We wish to test the hypothesis that using hair dyes *increases* the risk of breast cancer. We have

1. A one-tailed test with

$$\mathcal{H}_A : \pi > 7/1,000$$

2. Using the conventional choice of $\alpha = .05$ leads to the rejection region: $z \geq 1.65$
3. From the data,

$$p = \frac{20}{1,000}$$

$$= .02$$

and the hypothesis $\pi_0 = .007$,

$$z = \frac{.02 - .007}{\sqrt{\dfrac{(.007)(.993)}{1,000}}}$$

$$= 4.93$$

4. Since the computed z-score falls into the rejection region $(4.93 > 1.65)$, the null hypothesis is rejected at the chosen .05 level. In fact, the difference is very highly significant $(p < .001)$.

6.1.2. Matched Data

Here is the case where each subject or member of a group is observed twice for the presence or absence of a certain characteristic (e.g., at admission to and discharge from a hospital), or matched pairs are observed for the presence or absence of the same characteristic. A popular application is an epidemiologic design called a *pair-matched case–control study*. In case–control studies, cases of a specific disease are ascertained as they arise from population-based registers or lists of hospital admissions, and controls are sampled either as disease-free individuals from the population at risk or as hospitalized patients having a diagnosis other than the one under investigation. As a technique to control confounding factors, individual cases are matched, often one-to-one, to controls chosen to have similar values for confounding variables such as age, sex, race, and so forth. For pair-matched data with a single binary exposure (e.g., smoking vs. non-smoking), data can be represented by a 2×2 table where $(+, -)$ denotes (exposed, non-exposed) as in Table 6.1.

For example, b denotes the number of pairs where the case is exposed but the matched control is unexposed. The null hypothesis can be expressed either as

- The odds ratio associated with the exposure is 1.
- The cases and the controls have the same proportion of exposure.

TABLE 6.1 Pair-Matched Data With a Binary Exposure

Case	Control +	−
+	a	b
−	c	d

The decision is based on the standardized z-score

$$z = \frac{b-c}{\sqrt{b+c}}$$

or, in the two-tailed form, the square of the above statistic, denoted χ^2 (chi-squared)

$$\chi^2 = \frac{(b-c)^2}{b+c}$$

and the test is known as the *McNemar's chi-square*. If the test is one-tailed, z is used and the null hypothesis is rejected at the .05 level when

$$z \geq 1.65$$

If the test is two-tailed, χ^2 is used and the null hypothesis is rejected at the .05 level when

$$\chi^2 \geq 3.84$$

Example 6.2

It has been noted that metal workers have an increased risk for cancer of the internal nose and paranasal sinuses, perhaps as a result of exposure to cutting oils. Therefore, a study was conducted to see whether this particular exposure also increases the risk for squamous cell carcinoma of the scrotum.

Cases included all 45 squamous cell carcinomas of the scrotum diagnosed in Connecticut residents from 1955 to 1973, as obtained from the Connecticut Tumor Registry. Matched controls were selected for each case based on the age at death (within 8 years), year of death (within 3 years), and number of jobs as obtained from combined death certificate and Directory sources. An occupational indicator of metal worker (yes/no) was evaluated as the possible risk factor in this study; results are:

	Controls	
Case	Yes	No
Yes	2	26
No	5	12

We have, for a one-tailed test,

$$z = \frac{26 - 5}{\sqrt{26 + 5}}$$
$$= 3.77$$

indicating a very highly significant increase of risk associated with the exposure ($p < .001$).

Example 6.3

A study in Maryland identified 4,032 white persons, enumerated in a non-official 1963 census, who became widowed between 1963 and 1974. These people were matched, one-to-one, to married persons on the basis of race, sex, year of birth, and geography of residence. The matched pairs were followed to a second census in 1975, and we have the following overall male mortality:

Widowed men	Married men	
	Died	Alive
Died	2	292
Alive	210	700

An application of the McNemar's chi-square test (two-tailed) yields

$$\chi^2 = \frac{(292 - 210)^2}{292 + 210}$$
$$= 13.39$$

It can be seen that the null hypothesis of equal mortality should be rejected at the .05 level ($13.39 > 3.84$).

6.1.3. Comparison of Two or Several Independent Samples

Perhaps the most common problem involving proportions is the comparison of two independent samples.

In this type of problem, we have two independent samples of binary data (n_1, x_1) and (n_2, x_2) where the n's are adequately large sample sizes that may or may not be equal, the x's are the numbers of "positive" outcomes in the two samples, and we consider the null hypothesis

$$\mathcal{H}_0 : \pi_1 = \pi_2$$

expressing the equality of the two population proportions.

To perform a test of significance for $\mathcal{H}_0$, we proceed with the following steps:

1. Decide whether a one-tailed, say

$$\mathcal{H}_A : \pi_2 > \pi_1$$

or a two-tailed test,

$$\mathcal{H}_A : \pi_1 \neq \pi_2$$

is appropriate.
2. Choose a significance level α, a common choice being .05.
3. Calculate the z-score

$$z = \frac{p_2 - p_1}{\sqrt{p(1-p)\left(\dfrac{1}{n_1} + \dfrac{1}{n_2}\right)}}$$

where p is the "pooled proportion" defined by

$$p = \frac{x_1 + x_2}{n_1 + n_2}$$

4. Refer to the table for standard normal distribution (Appendix B) for selecting a cut point. For example, if the choice of α is .05, then the rejection region is determined by

- For the one-tailed alternative $\mathcal{H}_A : \pi_2 > \pi_1$, $z \geq 1.65$.
- For the one-tailed alternative $\mathcal{H}_A : \pi_2 < \pi_1$, $z \leq -1.65$.
- For the two-tailed test $\mathcal{H}_A : \pi_1 \neq \pi_2$, $z \leq -1.96$ or $z \geq 1.96$.

In the two-tailed form, the square of the z-score, denoted χ^2, is more often used. The test is referred to as the *chi-square test*. The test statistic can also be obtained using the short-cut formula

$$\chi^2 = \frac{(n_1 + n_2)[x_1(n_2 - x_2) - x_2(n_1 - x_1)]^2}{n_1 n_2 (x_1 + x_2)(n_1 + n_2 - x_1 - x_2)}$$

and the null hypothesis is rejected at the .05 level when

$$\chi^2 \geq 3.84$$

Example 6.4

A study was conducted to see whether an important public health intervention would significantly reduce the smoking rate among men. Of $n_1 = 100$ males sampled in 1965 at the time of the release of the Surgeon General's report on the health consequences of smoking, $x_1 = 51$ were found to be smokers. In 1980, a second random sample of $n_2 = 100$ males, similarly gathered, indicated that $x_2 = 43$ were smokers.

An application of the above method yields

$$p = \frac{51 + 43}{100 + 100}$$

$$= .47$$

$$z = \frac{.51 - .43}{\sqrt{(.47)(.53)\left(\dfrac{1}{100} + \dfrac{1}{100}\right)}}$$

$$= 1.13$$

It can be seen that the observed rate was reduced from 51% to 43%, but the reduction is not statistically significant at the .05 level ($z = 1.13 < 1.65$).

Example 6.5

An investigation was made into fatal poisonings of children by two drugs that were among the leading causes of such deaths. In each case, an inquiry was made as to how the child had received the fatal overdose and responsibility for the accident was assessed. Results were

	Drug A	Drug B
Child responsible	8	12
Child not responsible	31	19

We have the proportions of cases for which the child is responsible,

$$p_A = \frac{8}{8+31}$$
$$= .205 \text{ or } 20.5\%$$

$$p_B = \frac{12}{12+19}$$
$$= .387 \text{ or } 38.7\%$$

suggesting that they are not the same and that a child seems more prone to taking B than A. However, the chi-square statistic

$$\chi^2 = \frac{(39+31)[(8)(19)-(31)(12)]^2}{(39)(31)(20)(50)}$$
$$= 2.80 \qquad (< 3.84; \ \alpha = .05)$$

shows that the difference is not statistically significant at the .05 level.

Example 6.6

In Example 1.2, a case–control study was conducted to identify reasons for the exceptionally high rate of lung cancer among male residents of coastal Georgia. The primary risk factor under investigation was employment in shipyards during World War II, and the following table provides data for non-smokers:

Shipbuilding	Cases	Controls
Yes	11	35
No	50	203

We have for the cases

$$p_2 = 11/61$$
$$= .180$$

and for the controls

$$p_1 = 35/238$$
$$= .147$$

An application of the procedure yields a pooled proportion of

$$p = \frac{11 + 35}{61 + 238}$$
$$= .154$$

leading to

$$z = \frac{.180 - .147}{\sqrt{(.154)(.846)\left(\dfrac{1}{61} + \dfrac{1}{238}\right)}}$$
$$= .64$$

It can be seen that the rate of employment for the cases (18.0%) was higher than that for the controls (14.7%), but the difference is not statistically significant at the .05 level ($z = .64 < 1.65$).

There are situations where an investigator may want to adjust for a confounder that could influence the outcome of a statistical comparison. For example, in the above case–control study, it is reasonable to tabulate the data separately for different levels of smoking. The results are given in Table 6.2.

There are three 2×2 tables, one for each level of smoking; in Example 1.2, the last two tables were combined and presented together for simplicity.

Example 6.6 examines only data for non-smokers; however, it is more desirable to pool all the data together for a combined decision. The method, known as the *Mantel-Haenszel procedure,* is aimed to form a test statistic for each smoking group, then to combine the results into a single test statistic. We use the above data set and go through a step-by-step illustration of how the Mantel-Haenszel procedure is applied.

We begin with the 2×2 table for non-smokers (Table 6.3). New notations were added (r_1, r_2, c_1, c_2) to identify the row and column totals; the total frequency is

TABLE 6.2 Data for Different Levels of Smoking

Smoking	Shipbuilding	Cases	Controls
No	Yes	11	35
	No	50	203
Moderate	Yes	70	42
	No	217	220
Heavy	Yes	14	3
	No	96	50

TABLE 6.3 2 × 2 Table for Non-Smokers

Shipbuilding	Cases	Controls	Total
Yes	11 (a)	35	46 (r_1)
No	50	203	253 (r_2)
Total	61 (c_1)	238 (c_2)	299 (n)

n, and the frequency at the top left cell is denoted by a. The Mantel-Haenszel procedure is based on the z statistic

$$z = \frac{\sum a - \sum \frac{r_1 c_1}{n}}{\sqrt{\sum \frac{r_1 r_2 c_1 c_2}{n^2(n-1)}}}$$

where the summation ($\sum$) is across smoking levels. We have, for the non-smokers,

$$a = 11$$

$$\frac{r_1 c_1}{n} = \frac{(46)(61)}{299}$$

$$= 9.38$$

$$\frac{r_1 r_2 c_1 c_2}{n^2(n-1)} = \frac{(46)(253)(61)(238)}{(299)^2(298)}$$

$$= 6.34$$

The process is repeated for each of the other two smoking levels. For moderate smokers

$$a = 70$$

$$\frac{r_1 c_1}{n} = \frac{(112)(287)}{549}$$

$$= 58.55$$

$$\frac{r_1 r_2 c_1 c_2}{n^2(n-1)} = \frac{(112)(437)(287)(262)}{(549)^2(548)}$$

$$= 22.28$$

and for heavy smokers

$$a = 14$$

$$\frac{r_1 c_1}{n} = \frac{(17)(110)}{163}$$

$$= 11.47$$

$$\frac{r_1 r_2 c_1 c_2}{n^2(n-1)} = \frac{(17)(146)(110)(53)}{(163)^2(162)}$$

$$= 3.36$$

These results are combined to obtain the z-score

$$z = \frac{(11 - 9.38) + (70 - 58.55) + (14 - 11.47)}{\sqrt{6.34 + 22.28 + 3.36}}$$

$$= 2.76$$

and a z-score of 2.76 yields a one-tailed p-value of .0029, which is beyond the 1% level. This result is stronger than that of the previous example because it is based on more information where all data at all three smoking levels are used. For a two-tailed decision, use $\chi^2 = z^2$ and the null hypothesis of no difference is rejected when $\chi^2 \geq 3.84$.

Example 6.7

A case–control study was conducted to investigate the relationship between myocardial infarction (MI) and oral contraceptive use (OC). The data, stratified by cigarette smoking, were

Smoking	OC use	Cases	Controls
No	Yes	4	52
	No	34	754
Yes	Yes	25	83
	No	171	853

An application of the Mantel-Haenszel procedure yields

1.

	Smoking	
	No	Yes
a	4	25
$\dfrac{r_1 c_1}{n}$	2.52	18.70
$\dfrac{r_1 r_2 c_1 c_2}{n^2(n-1)}$	2.25	14.00

2. The combined z-score is

$$z = \frac{(4-2.52)+(25-18.70)}{\sqrt{2.25+14.00}}$$
$$= 1.93$$

which is significant at the 5% level (one-tailed).

Consider now the problem we would face if there were more than two groups to compare. For example, Exercise 6.5 presents a case–control study of esophageal cancer among men; the aim was to determine whether alcohol consumption is a risk factor. Since both cancer incidence rate and alcohol consumption rate may rise with age, age must always be considered a confounder. Data from the 785 controls can be listed as in Table 6.4.

Suppose we want to see whether the three age groups have the same daily alcohol consumption rate. Of course, we could ask questions such as, "Do the 25–44 and

TABLE 6.4   Possible Association Between Alcohol Consumption and Age

Daily alcohol consumption	Age groups		
	25–44	45–64	65+
80+ g	35	56	18
0–79 g	270	277	129
Total	305	333	147

**TABLE 6.5   Data To Determine if the
25–44 and 45–64 Year Age Groups Differ**

Daily alcohol consumption	Age groups		
	25–44	45–64	Total
80+ g	35	56	91
0–79 g	270	277	547
Total	305	333	638

45–64 age groups have the same rate?" "Do the 25–44 and 65+ age groups have the same rate?" "Do the 45–64 and 65+ age groups have the same rate?" In each case, we would have two proportions to compare, and we would have to perform the chi-square test of Section 6.1.3. For example, to answer the first question, we use the data set shown in Table 6.5. An application of the chi-square test yields

$$\chi^2 = \frac{638[(35)(277) - (270)(56)]^2}{(305)(333)(91)(547)}$$

$$= 3.71 \quad (\text{vs. } 3.84 \text{ at } \alpha = .05)$$

indicating that the difference between the two age groups is almost significant at the .05 level, the rates being 11.5% for the 25–44 age group and 16.8% for the 45–64 age group. The other two questions may be answered similarly.

What is the matter with this approach of doing many chi-square tests, one for each pair of samples? As the number of groups increases, so does the number of tests to perform; for example, we would have to do 45 tests if we have 10 groups to compare. Obviously, the amount of work is greater, but that should not be the critical problem—especially with technologic aids such as calculators and computers. So, what is the problem? The answer is that performing many tests increases the probability that one or more of the comparisons will result in a Type I error (i.e., a significant test result when the null hypothesis is true). This statement should make sense intuitively. For example, suppose the null hypothesis is true and we perform 100 tests—each has a .05 probability of resulting in a Type I error; then 5 of these 100 tests would result in Type I errors. Of course, we usually do not need to do that many tests; however, every time we do more than one, then the probability that at least one will result in a Type I error exceeds .05, indicating a falsely significant difference!

What is required is a method of simultaneously comparing several proportions in one step. To form this procedure, let us put the data in a general form without considering any specific example. In this type of problem, we have k independent samples of binary data $(n_1, x_1), (n_2, x_2), \ldots, (n_k, x_k)$, where the n's are sample sizes and the x's are the numbers of positive outcomes in the k samples.

Let

$$p_i = x_i / n_i; \quad i = 1, 2, \ldots, k$$

be the sample proportion of group i. We consider the null hypothesis

$$\mathcal{H}_0 : \pi_1 = \pi_2 = \cdots = \pi_k$$

expressing the equality of the k population proportions. In the example introduced at the beginning of this section, we have

$$k = 3$$

$$n_1 = 305, \qquad x_1 = 35$$

$$p_1 = 35/305$$

$$= .115$$

$$n_2 = 333, \qquad x_2 = 56$$

$$p_2 = 56/333$$

$$= .168$$

$$n_3 = 147, \qquad x_3 = 18$$

$$p_3 = 18/147$$

$$= .122$$

Let p be the "pooled proportion" defined by

$$p = \frac{x_1 + x_2 + \cdots + x_k}{n_1 + n_2 + \cdots + n_k}$$

The test statistic is given by the following formula:

$$\chi^2 = \frac{\sum n_i (p_i - p)^2}{p(1 - p)}$$

where the summation is over the k groups and the decision is made referring to the chi-square table (Appendix D) at the .05 level and $(k - 1)$ degrees of freedom. With the above example, we have

$$p = \frac{35 + 56 + 18}{305 + 333 + 147}$$

$$= \frac{109}{785}$$

$$= .139$$

leading to

$$\chi^2 = \frac{305(.115 - .139)^2 + 333(.168 - .139)^2 + 147(.122 - .139)^2}{(.139)(1 - .139)}$$

$$= 4.163$$

The difference between the three groups is not significant at the .05 level because the cut point for chi-square with 2 degrees of freedom is 5.991.

Let us suppose we want to apply this procedure to the comparison of two proportions. Then it can be seen that we would have the same result as if we were to use the method of Section 6.1.3. In other words, the method of this section is a natural extension of the chi-square test from the comparison of two proportions to the comparison of several proportions.

Example 6.8

Some investigators were interested in studying changing patterns in soft-tissue sarcomas over time. There are three principle types of these sarcomas, one of which is *fibroid*, which is characterized by a muscle cell origin. To study this question the investigators utilized data on soft-tissue sarcomas of the arms and legs from the Connecticut Tumor Registry; part of the data are tabulated as follows.

	Decade			
Tissue type	**1935–44**	**1945–54**	**1955–64**	**Total**
Fibroid	40	70	93	203
Others	33	42	85	160
Total	73	112	178	363

An application of the above chi-square test yields

$$k = 3$$
$$n_1 = 73, \qquad x_1 = 40$$
$$p_1 = 40/73$$
$$\quad = .548$$
$$n_2 = 112, \qquad x_2 = 70$$
$$p_2 = 70/112$$
$$\quad = .625$$
$$n_3 = 178, \qquad x_3 = 93$$
$$p_3 = 93/178$$
$$\quad = .522$$

and

$$p = 203/363$$

$$= .559$$

leading to

$$\chi^2 = \frac{73(.548 - .559)^2 + 112(.625 - .559)^2 + 178(.522 - .559)^2}{(.559)(1 - .559)}$$

$$= 3.003$$

The difference between the three decades is not significant because the cut point for chi-square with two degrees of freedom is 5.991.

The problem in Example 6.8 involves the comparison of several proportions; there are several groups, and the subjects in each group are classified into two categories. The same test procedure is also applicable to problems where we wish to compare two groups in which each subject could be classified into one of several categories. The example on alcohol consumption at the beginning of this section serves as one example; the following is another.

Example 6.9

In the course of selecting controls for a study to evaluate the effect of caffeine-containing coffee on the risk of myocardial infarction among women 30–49 years of age, a study noted appreciable differences in coffee consumption among hospital patients admitted for illnesses not known to be related to coffee use. Among potential controls, the coffee consumption of patients who had been compelled to hospital by conditions having an acute onset (such as fractures) was compared to that of patients admitted for chronic disorders.

	Cups of coffee per day			
Group	**0**	**1–4**	**≥ 5**	**Total**
Acute conditions	340	457	183	980
Chronic conditions	2,440	2,527	868	5,835
Total	2,780	2,984	1,051	6,815

An application of the above chi-square test yields

$$k = 3$$
$$n_1 = 2{,}780, \qquad x_1 = 340$$
$$p_1 = 340/2{,}780$$
$$= .122$$
$$n_2 = 2{,}984, \qquad x_2 = 457$$
$$p_2 = 457/2{,}984$$
$$= .153$$
$$n_3 = 1{,}051, \qquad x_3 = 183$$
$$p_3 = 183/1{,}051$$
$$= .174$$

and

$$p = 980/6{,}815$$
$$= .142$$

leading to

$$\chi^2 = \frac{2{,}780(.122 - .142)^2 + 2{,}984(.153 - .142)^2 + 1{,}051(.174 - .142)^2}{(.142)(1 - .142)}$$
$$= 20.958$$

The difference between the two groups, those with acute conditions and those with chronic conditions, has been shown to be highly significant, the cut point for chi-square with 2 degrees of freedom $(2 = k - 1)$ being 5.991.

6.2. COMPARISONS OF MEANS

6.2.1. One-Sample Problem

In the one-sample problem, we have a sample of continuous measurements of size n and we consider the null hypothesis

$$\mathcal{H}_0 : \mu = \mu_0$$

where μ_0 is a fixed and known number. It is often a standardized or referenced figure.

To perform a test of significance for $\mathcal{H}_0$, we proceed with the following steps:

1. Decide whether a one-tailed or a two-tailed test is appropriate; this decision depends on the research question.
2. Choose a level of significance; a common choice is .05.
3. Calculate the t-statistic

$$t = \frac{\bar{x} - \mu_0}{SE(\bar{x})}$$

4. From the table for t-distribution (Appendix C) with $(n-1)$ degrees of freedom and the choice of α, e.g., $\alpha = .05$, the rejection region is determined by

   - For a one-tailed test, use the column corresponding to the "upper tail area" of .05:

$$t \leq -\text{tabulated value for } \mathcal{H}_A : \mu < \mu_0$$

$$t \geq \text{tabulated value for } \mathcal{H}_A : \mu > \mu_0$$

   - For a two-tailed test or $\mathcal{H}_A : \mu \neq \mu_0$, use the column corresponding to the "upper tail area" of .025

$$z \leq -\text{tabulated value or } z \geq \text{tabulated value}$$

Example 6.10

Boys of a certain age have a mean weight of 85 lbs. An observation was made that in a city neighborhood, children were underfed. As evidence, all 25 boys in the neighborhood of that age were weighed and found to have a mean $\bar{x}$ of 80.94 lbs and a standard deviation s of 11.60 lbs. An application of the above procedure yields

$$SE(\bar{x}) = \frac{s}{\sqrt{n}}$$

$$= \frac{11.60}{\sqrt{25}}$$

$$= 2.32$$

leading to

$$t = \frac{80.94 - 85}{2.32}$$

$$= -1.75$$

The underfeeding complaint corresponds to the one-tailed alternative

$$\mathcal{H}_A : \mu < 85$$

so that we would reject the null hypothesis if

$$t \leq -\text{tabulated value}$$

From Appendix C, and with 24 degrees of freedom $(n-1)$, we find

$$\text{Tabulated value} = 1.71$$

under the column corresponding to .05 upper tail area; the null hypothesis is rejected at the .05 level. In other words, there is enough evidence to support the underfeeding complaint.

6.2.2. Matched Data

For the comparison of means, we distinguish between two situations. Sometimes we take two independent samples, but in many other investigations the experimental group also serves as its own control. As indicated in Section 4.2.3, data from matched or before-and-after experiments should never be considered as coming from two independent samples. The procedure is to reduce the data to a one-sample by computing before-and-after (or case-and-control) differences for each subject or pairs of matched subjects. By doing this with paired observations, we get a set of differences that can be handled as a one-sample problem. In this case the null hypothesis to be considered is

$$\mathcal{H}_0 : \mu_d = 0$$

a special case of the method in the previous section (μ_d is the "mean of the differences"). The test statistic is

$$t = \frac{\overline{d}}{\text{SE}(\overline{d})}$$

and the rejection region is determined using the t-distribution at $(n-1)$ degrees of freedom. This test is referred to as the *one-sample t-test*.

Example 6.11

The systolic blood pressures of $n = 12$ women between the ages of 20 and 35 years were measured before and after administration of a newly developed oral contraceptive. Raw data were tabulated in Example 4.5 in Section 4.2.3, and the following are the necessary summarized figures:

$$\bar{d} = 2.58 \text{ mmHg}$$

and

$$s = 3.09$$

An application of the above procedure yields

$$SE(\bar{d}) = \frac{s}{\sqrt{n}}$$

$$= \frac{3.09}{\sqrt{12}}$$

$$= .89$$

$$t = \frac{2.58}{.89}$$

$$= 2.90$$

Using the column corresponding to the upper tail area of .05 in Appendix C, we have a tabulated value of 1.796 for 11 degrees of freedom. Since

$$t = 2.90 > 2.201$$

we conclude that the null hypothesis of no blood pressure change should be rejected at the .05 level; there is enough evidence to support the hypothesis of increased systolic blood pressure (one-tailed alternative).

6.2.3. Comparison of Two Independent Samples

For the case of two independent samples, the test statistic is

$$t = \frac{\bar{x}_1 - \bar{x}_2}{SE(\bar{x}_1 - \bar{x}_2)}$$

where

$$SE(\bar{x}_1 - \bar{x}_2) = s_p \sqrt{\frac{1}{n_1} + \frac{1}{n_2}}$$

$$s_p^2 = \frac{(n_1 - 1)s_1^2 + (n_2 - 1)s_2^2}{n_1 + n_2 - 2}$$

This test is referred to as the *two-sample t-test*, and its rejection region is determined using the *t*-distribution at $(n_1 + n_2 - 2)$ degrees of freedom:

- For a one-tailed test, use the column corresponding to the "upper tail area" of .05, and $\mathcal{H}_0$ is rejected if

$$t \leq -\text{tabulated value for } \mathcal{H}_A : \mu_1 < \mu_2$$

or

$$t \geq \text{tabulated value for } \mathcal{H}_A : \mu_1 > \mu_2$$

- For a two-tailed test or $\mathcal{H}_A : \mu_1 \neq \mu_2$, use the column corresponding to the "upper tail area" of .025, and $\mathcal{H}_0$ is rejected if

$$t \leq -\text{tabulated value or } t \geq \text{tabulated value}$$

Example 6.12

In an attempt to assess the physical condition of joggers, a sample of $n_1 = 25$ joggers was selected and their maximum volume of oxygen uptake (VO_2) were measured with the following results:

$$\bar{x}_1 = 47.5 \text{ ml/kg}, \qquad s_1 = 4.8 \text{ ml/kg}$$

Results for a sample of $n_2 = 26$ non-joggers were

$$\bar{x}_2 = 37.5 \text{ ml/kg}, \qquad s_2 = 5.1 \text{ ml/kg}$$

To proceed with the two-tailed, two-sample *t*-test, we have

$$s_p^2 = \frac{(24)(4.8)^2 + (25)(5.1)^2}{49}$$

$$= 24.56$$

$$\text{or} \qquad s_p = 4.96$$

$$\text{SE}(\bar{x}_1 - \bar{x}_2) = 4.96\sqrt{\frac{1}{25} + \frac{1}{26}}$$

$$= 1.39$$

It follows that

$$t = \frac{47.5 - 37.5}{1.39}$$

$$= 7.19$$

indicating a significant difference between the joggers and the non-joggers (at 49 degrees of freedom, the tabulated value is about 2.0).

Example 6.13

Vision, or, more specially, visual acuity, depends on a number of factors. A study was undertaken in Australia to determine the effect of one of these factors: racial variation. Visual acuity of recognition as assessed in clinical practice has a defined normal value of 20/20 (or zero in log scale). The following summarized data on monocular visual acuity (expressed in log scale) were obtained from two groups:

1. Australian males of European origin

$$n_1 = 89$$

$$\bar{x}_1 = -.20$$

$$s_1 = .18$$

2. Australian males of Aboriginal origin

$$n_2 = 107$$

$$\bar{x}_2 = -.26$$

$$s_2 = .13$$

To proceed with a two-sample t-test we have

$$s_p^2 = \frac{(88)(.18)^2 + (106)(.13)^2}{194}$$

$$= (.155)^2$$

$$SE(\bar{x}_1 - \bar{x}_2) = (.155)\sqrt{\frac{1}{89} + \frac{1}{107}}$$

$$= .022$$

$$t = \frac{(-.20) - (-.26)}{.022}$$

$$= 2.73$$

The result indicates that the difference is statistically significant beyond the .01 level (at $\alpha = .01$ and for a two-tailed test the cut point being 2.58 for large degrees of freedom).

Suppose that the goal of a research project is to discover whether there are differences in the means of several independent groups. The problem is how we will measure the extent of differences among the means. If we had two groups, then we would measure the difference by the distance between sample means $(\bar{x} - \bar{y})$ and use the two-sample t-test of Section 6.2.3. Here we have more than two groups; we could take all possible pairs of means and do many two-sample t-tests. However, remember from the section on the chi-square test that making multiple tests increases the probability of making a Type I error. What is needed is a single measure that summarizes the differences between several means and a method of simultaneously comparing these means in one step. This method is called ANOVA, which is an abbreviation of "ANalysis Of VAriance." This useful method is not presented here because it is beyond the scope of this book.

6.3. REGRESSION AND CORRELATION

Regression and correlation form a pair of topics, closely connected to each other. Regression, although the colloquial meaning of the word sounds "negative" as though something is sliding backwards or getting worse, is one of the most useful concepts and techniques of statistics. Correlation, both a concept and a technique, has a technical meaning different enough from the colloquial meaning to cause serious confusion. Both regression and correlation are so important, even at the elementary level, that we've decided to give them some coverage in this text.

Statistical regression, as a technique, is really neat and clever. How many times have you looked at a bunch of points that sort of lie on a line and tried to draw a straight line through them? Statistical regression provides a way to draw that line and even gives a sound theoretical reason for doing it that way. Without much additional effort it will even draw curves to fit scattered data. You even get help in interpreting the lines or curves that the formulas tell you (or your computer) to draw.

Correlation is a concept, with common colloquial usage of association, such as "height and weight are correlated." Statisticians have given a technical meaning to it; they can actually calculate a number that tells the *amount* of association.

The number is a slippery one to interpret, however. A correlation number can't be smaller than −1 or larger than +1. It is true that a scatter plot of data that results in a correlation number of +1 or −1 has to lie in a perfectly straight line, but hardly any real data are like that. A correlation of 0, however, doesn't mean that there is no association. It means there is no *linear* association. You can have a correlation near 0, and yet have a very useful association, such as the case when the data fall neatly on a sharply bending curve. It takes a lot of skill to learn how to interpret correlation numbers such as .3 or .62. Such numbers can be statistically significant or not, or be useful or not. We advise getting expert statistical help when interpreting correlation, especially when a scatter plot (graph) of the data is not available.

6.3.1. Basic Concepts

Section 6.2 is mainly concerned with a single continuous measurement made on each element of a sample. However, in many important investigations we may have two measurements made, that is, where the sample consists of pairs of values and the research objective is concerned with the association between these variables. For example, what is the relationship between a mother's weight and her baby's weight?

Sections 4.3.3 and 6.1 are concerned with the association between dichotomous variables. For example, if we want to investigate the relationship between a disease and a certain risk factor, we could undertake a case–control study and present the data in the form of a 2×2 table as shown in Table 6.6. Using Table 6.6 we could

1. Perform a hypothesis test to compare the characteristics of exposure for the cases versus the controls to determine if a relationship exists.
2. Calculate an odds ratio to represent the strength of the relationship.

TABLE 6.6 2×2 Table for Case–Control Study

Smoking	Lung cancer	
	Cases	Controls
Yes	a	b
No	c	d

This section deals with continuous measurements, and the method is referred to as *regression and correlation analysis*.

When dealing with the relationship between two continuous variables, we first have to distinguish between a *deterministic* relationship and a *statistical* relationship. For a deterministic relationship, values of the two variables are related through an exact mathematical formula. For example, consider the relationship between hospital cost and number of days in hospital. If the costs are $100 for admission and $150 per day, then we can easily calculate the total cost given the number of days in hospital, and, if any set of data is plotted, say, cost versus number of days, all data points fall perfectly on a straight line. A statistical relationship, unlike a deterministic one, is not a perfect one. In general, the points do not fall perfectly on any line or curve. For example, Table 6.7 gives the values for 12 births of the birth weight (x) and the increase in weight between days 70 and 100 of life, expressed as a percentage of the birth weight (y). If we let each pair of numbers (x, y) be represented by a dot in a diagram with the x's on the horizontal axis, we obtain Figure 6.1.

The dots do not fall perfectly on a straight line, but rather scatter around one, very typical for statistical relationships. Because of this scattering of dots, the diagram is called a *scatter diagram*. The positions of the dots provide some information about the direction as well as the strength of the association under the investigation. If they tend to go from lower left to upper right, we have a positive association; if they tend to go from upper left to lower right, we have a negative association. The relationship becomes weaker and weaker as the distribution of the dots clusters less closely around the line and becomes virtually no correlation when the distribution approximates a circle or oval (the method is ineffective for measuring a relationship that is not linear).

TABLE 6.7 Birth Weight Data

x (oz)	y (%)
112	63
111	66
107	72
119	52
92	75
80	118
81	120
84	114
118	42
106	72
103	90
94	91

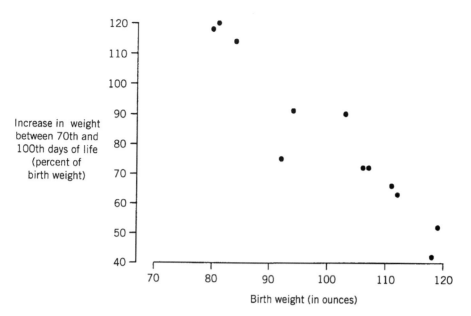

Fig. 6.1 Scatter diagram for birth weight data.

6.3.2. Coefficient of Correlation

Consider again a scatter diagram, shown below, where we added a vertical and a horizontal line through the point $(\bar{x}, \bar{y})$ and label the four quarters as I, II, III, and IV (Figure 6.2).

It can be seen that

- In quarters I and III,

$$(x - \bar{x})(y - \bar{y}) > 0$$

so that for positive association, we have

$$\sum (x - \bar{x})(y - \bar{y}) > 0$$

Furthermore, this sum is large for stronger relationships because most of the dots, being closely clustered around the line, are in these two quarters.

- Similarly, in quarters II and IV,

$$(x - \bar{x})(y - \bar{y}) < 0$$

leading to

$$\sum (x - \bar{x})(y - \bar{y}) < 0$$

for negative association.

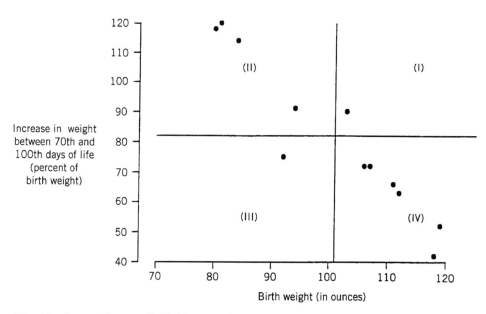

Fig. 6.2 Scatter diagram divided into quadrants.

With a proper standardization, we obtain

$$r = \frac{\sum(x - \bar{x})(y - \bar{y})}{\sqrt{\left[\sum(x - \bar{x})^2\right]\left[\sum(y - \bar{y})^2\right]}}$$

so that

$$-1 \le r \le 1$$

This statistic r is called the *correlation coefficient* and is a popular measure for the strength of a statistical relationship.

- Values near 1 indicate a strong positive association.
- Values near -1 indicate a strong negative association.
- Values around 0 indicate a weak association.

In computing we more often use

$$r = \frac{\sum xy - \frac{(\sum x)(\sum y)}{n}}{\sqrt{\left[\sum x^2 - \frac{(\sum x)^2}{n}\right]\left[\sum y^2 - \frac{(\sum y)^2}{n}\right]}}$$

Example 6.14

Consider again the previous birth weight problem. We have

	x	y	x^2	y^2	xy
	112	63	12,544	3,969	7,056
	111	66	12,321	4,356	7,326
	107	72	11,449	5,184	7,704
	119	52	14,161	2,704	6,188
	92	75	8,464	5,625	6,900
	80	118	6,400	13,924	9,440
	81	120	6,561	14,400	9,720
	84	114	7,056	12,996	9,576
	118	42	13,924	1,764	4,956
	106	72	11,236	5,184	7,632
	103	90	10,609	8,100	9,270
	94	91	8,836	8,281	8,554
Totals	1,207	975	123,561	86,487	94,322

Using these five totals, we obtain

$$r = \frac{94,322 - \dfrac{(1,207)(975)}{12}}{\sqrt{\left[123,561 - \dfrac{(1,207)^2}{12}\right]\left[86,487 - \dfrac{(975)^2}{12}\right]}}$$

$$= -.946$$

indicating a very strong negative association.

Example 6.15

The following data represent systolic blood pressure readings on 15 women:

Age (x)	SBP (y)	Age (x)	SBP (y)
42	130	85	162
46	115	72	158
42	148	64	155
71	100	81	160
80	156	41	125
74	162	61	150
70	151	75	165
80	156		

We set up a work table as in Example 6.14:

	x	y	x^2	y^2	xy
	42	130	1,764	16,900	5,460
	46	115	2,116	13,225	5,290
	42	148	1,764	21,904	6,216
	71	100	5,041	10,000	7,100
	80	156	6,400	24,336	12,480
	74	162	5,476	26,224	11,988
	70	151	4,900	22,801	10,570
	80	156	6,400	24,336	12,480
	85	162	7,225	26,224	13,770
	72	158	5,184	24,964	11,376
	64	155	4,096	24,025	9,920
	81	160	6,561	25,600	12,960
	41	125	1,681	15,625	5,125
	61	150	3,721	22,500	9,150
	75	165	5,625	27,225	12,375
Totals	984	2,193	67,954	325,889	146,260

Using these totals, we obtain

$$r = \frac{146{,}260 - \dfrac{(984)(2{,}193)}{15}}{\sqrt{\left[67{,}954 - \dfrac{(984)^2}{15}\right]\left[325{,}889 - \dfrac{(2{,}193)^2}{15}\right]}}$$

$$= .566$$

indicating a moderately positive association.

6.3.3. Testing for Independence

The correlation coefficient r describes the relationship between the sample observations for two variables. It is an estimate of an unknown population correlation coefficient ρ (rho) the same way the sample mean $\bar{x}$ is used as an estimate of some unknown population mean μ. We are usually interested in knowing if we may conclude that $\rho \neq 0$, that is, that the two variables under investigation are really correlated. The test statistic is

$$t = r\sqrt{\frac{n-2}{1-r^2}}$$

The procedure is often performed as two-tailed, that is,

$$\mathcal{H}_A : \rho \neq 0$$

and it is a t-test with $(n-2)$ degrees of freedom.

Example 6.16

For the birth weight problem of Example 6.14, we have

$$n = 12$$

$$r = -.946$$

leading to

$$t = (-.946)\sqrt{\frac{10}{1-(-.946)^2}}$$

$$= -9.23$$

At $\alpha = .05$ and df $= 10$ degrees of freedom, the tabulated t-coefficient is 2.228, indicating that the null hypothesis of independence should be rejected

$$t = -9.23 \quad (t < -2.228).$$

Example 6.17

For the blood pressure problem of Example 6.15, we have

$$n = 15$$

$$r = .566$$

leading to

$$t = (.566)\sqrt{\frac{13}{1 - (.566)^2}}$$

$$= 2.475$$

At $\alpha = .05$ and df = 13 degrees of freedom, the t tabulated value is 2.16. Since

$$t > 2.16$$

we have to conclude that the null hypothesis of independence should be rejected; that is, the relationship between age and systolic blood pressure is real.

In cases where X and Y are not independent, we can use *regression analysis* to help us predict the value of Y (called the *dependent variable*) that is associated with a given value of X (called the *independent variable*). In most situations involving regression data, the points do not all lie on a straight line; however, we are always able to construct a *regression line* that best fits the data. You will recall from beginning algebra that a straight line placed on a set of coordinates has a slope (b) and a Y-intercept (a), and they are part of the general equation

$$Y = a + bX$$

You should also recall that, if we know the slope b and the Y-intercept a, it would be fairly easy to plot the line on graph paper and that "$a + bX$" is the value of Y that is associated with the given value X. For regression data, the line that fits the dots (by the method called *least squares*) has its slope given by

$$b = \frac{\sum xy - \dfrac{(\sum x)(\sum y)}{n}}{\sum x^2 - \dfrac{(\sum x)^2}{n}}$$

and Y-intercept by

$$a = \bar{y} - b\bar{x}$$

Example 6.18

For the birth weight problem of Example 6.14, we have

$$b = \frac{94{,}322 - \dfrac{(1{,}207)(975)}{12}}{123{,}561 - \dfrac{(1{,}207)^2}{12}}$$

$$= -1.74$$

$$\bar{x} = 1{,}207/12$$

$$= 100.6$$

$$\bar{y} = 975/12$$

$$= 81.3$$

$$a = 81.3 - (-1.74)(100.6)$$

$$= 256.3$$

For example, if the birth weight is 95 ounces, it is predicted that the increase between days 70 and 100 of life would be

$$y = 256.3 + (-1.74)(95)$$

$$= 90.1\% \text{ of birth weight}$$

Example 6.19

For the blood pressure problem of Example 6.15, we have

$$b = \frac{146{,}260 - \dfrac{(984)(2{,}193)}{15}}{67{,}954 - \dfrac{(984)^2}{15}}$$

$$= .71$$

$$\bar{x} = 984/15$$

$$= 65.6$$

$$\bar{y} = 2{,}193/15$$

$$= 146.2$$

$$a = 146.2 - (.71)(65.6)$$

$$= 99.6$$

For example, for a 50-year-old woman, it is predicted that her systolic blood pressure would be about

$$y = 99.6 + (.71)(50)$$

$$= 135 \text{ mmHg}$$

EXERCISES

1. Consider a sample of $n = 110$ women drawn randomly from the membership list of the National Organization for Women (NOW), $x = 25$ of whom were found to smoke. Use the result of this sample to test whether the rate found is significantly *different* from the U.S. proportion of .30 for women.

2. A matched case–control study on endometrial cancer, where the exposure was "ever having taken any estrogen," yields the following data:

Case	Matching control	
	Exposed	Non-exposed
Exposed	27	29
Non-exposed	3	4

Compare the cases versus the controls.

3. Ninety-eight heterosexual couples, at least one of whom was HIV-infected, were enrolled in an HIV transmission study and interviewed about sexual behavior.

The following table provides a summary of condom use reported by heterosexual partners:

Woman	Man		Total
	Ever	Never	
Ever	45	6	51
Never	7	40	47
Total	52	46	98

Compare the men and the women; state your null hypothesis.

4. A matched case–control study was conducted to evaluate the cumulative effects of acrylate and methacrylate vapors on olfactory function. Cases were defined as scoring at or below the 10th percentile on the UPSIT (University of Pennsylvania Smell Identification Test).

Controls	Cases	
	Exposed	Unexposed
Exposed	25	22
Unexposed	9	21

Compare the cases versus the controls; state your null hypothesis.

5. Epidemic keratoconjunctivitis (EKC), or "shipyard eye," is an acute infectious disease of the eye. A case of EKC is defined as an illness

- Consisting of redness, tearing, and pain in one or both eyes for more than 3 days duration
- Diagnosed as EKC by an ophthalmologist

In late October 1977, one (Physician A) of the two ophthalmologists providing the majority of specialized eye care to the residents of a central Georgia county (population 45,000) saw a 27-year-old nurse who had returned from a vacation in Korea with severe EKC. She received symptomatic therapy and was warned that her eye infection could spread to others; nevertheless, numerous cases of an illness similar to hers soon occurred in the patients and staff of the nursing home (Nursing Home A) where she worked (these individuals came to Physician A for diagnosis and treatment). The following table provides exposure history of 22 persons with EKC between October 27, 1977, and January 13, 1978 (when the outbreak stopped after proper control techniques were initiated).

Nursing Home B, included in this table, is the only other area chronic-care facility.

Exposure cohort	No. exposed	No. of cases
Nursing home A	64	16
Nursing home B	238	6

Using an appropriate test, compare the proportions of cases from the two nursing homes.

6. In August 1976, tuberculosis was diagnosed in a high school student (index case) in Corinth, Mississippi. Subsequently, laboratory studies revealed that the student's disease was caused by drug-resistant tubercule bacilli. An epidemiologic investigation was conducted at the high school.

   The following table gives the rate of positive tuberculin reactions, determined for various groups of students according to degree of exposure to the index case.

Exposure level	No. tested	No. positive
High	129	63
Low	325	36

Compare the proportions of positive cases for the two exposure levels using an appropriate test of significance.

7. Consider the data taken from a study that attempts to determine whether the use of electronic fetal monitoring (EFM) during labor affects the frequency of cesarean section deliveries. Of the 5,824 infants included in the study, 2,850 were electronically monitored and 2,974 were not. The outcomes are as follows:

Cesarean delivery	EFM exposure		Total
	Yes	No	
Yes	358	229	587
No	2,492	2,745	5,237
Total	2,850	2,974	5,824

Compare the proportions of cesarean delivery for the two exposure groups.

8. A study was conducted to investigate the effectiveness of bicycle safety helmets in preventing head injury. The data consist of a random sample of 793 individuals who were involved in bicycle accidents during a 1-year period.

	Wearing helmet		
Head injury	Yes	No	Total
Yes	17	218	235
No	130	428	558
Total	147	646	793

Compare the proportions of head injury for the group with helmets versus the group without helmets.

9. Since incidence rates of most cancers rise with age, age must always be considered a confounder. The following are stratified data for an unmatched case–control study. The disease was esophageal cancer among men, and the risk factor was alcohol consumption.

		Daily alcohol consumption	
Age		80+ g	0–79 g
25–44	Cases	5	5
	Controls	35	270
45–64	Cases	67	55
	Controls	56	277
65+	Cases	24	44
	Controls	18	129

Use the Mantel-Haenszel procedure to compare the cases versus the controls.

10. Prematurity, which ranks as the major cause of neonatal morbidity and mortality, has traditionally been defined on the basis of a birth weight under 2,500 g. However, this definition encompasses at least two distinct types of infants: infants who are small because they are born early, and infants who are born at or near term but are small because their growth was retarded. The term *prematurity* has now been replaced by

- *low birth weight*, to describe the second type
- *preterm*, to characterize the first type, i.e., a baby born before 37 weeks of gestation

A case–control study of the epidemiology of preterm delivery was undertaken at Yale–New Haven Hospital in Connecticut during 1977. The study population consisted of 175 mothers of preterm infants and 303 mothers of full-term infants. The following table gives the distribution of socioeconomic status:

Status	Cases	Controls
Middle-upper	58	148
Lower	117	155

What conclusion do you draw?

11. Risk factors of gallstone disease were investigated in male self-defense officials who received, between October 1986 and December 1990, a retirement health examination at the Self-Defense Forces Fukuoka Hospital, Fukuoka, Japan. The following are parts of the data:

Factor	No. of men surveyed	
	Total	No. with gallstones
Smoking		
Never	621	11
Past	776	17
Current	1,342	33
Alcohol		
Never	447	11
Past	113	3
Current	2,179	47
Body mass index (kg/m^2)		
< 22.5	719	13
22.5–24.9	1,301	30
≥ 25.0	719	18

Arrange the data into 2×3 tables and use the chi-square test to determine the effects of each factor.

12. A case–control study was conducted in Auckland, New Zealand, to investigate the effects of alcohol consumption on both nonfatal myocardial infarction and coronary death in the 24 hours after drinking among regular drinkers. Data were tabulated separately for men and women.

Drink in the last 24 hours	Myocardian infarction		Coronary death	
	Controls	Cases	Controls	Cases
Men				
No	197	142	135	103
Yes	201	136	159	69
Women				
No	144	41	89	12
Yes	122	19	76	4

Determine the effects on each event of each group, and calculate the 95% confidence interval for the corresponding odds ratio.

13. The same study referenced in Example 6.3 also provided mortality data for 2,828 matched pairs of women.

Widowed women	Married women	
	Died	Alive
Died	1	264
Alive	249	2,314

Test to see whether the null hypothesis of equal mortality should be rejected.

14. Postmenopausal women who develop endometrial cancer are on the whole heavier than women who do not develop the disease. One possible explanation is that heavy women are more exposed to endogenous estrogens that are produced in postmenopausal women by conversion of steroid precursors to active estrogens in peripheral fat. In the face of varying levels of endogenous estrogen production, one might ask whether the carcinogenic potential of exogenous estrogens would be the same in all women. A study has been conducted to examine the relation between weight, replacement estrogen therapy, and endometrial cancer in a case–control study.

Weight (kg)		Estrogen replacement	
		Yes	No
< 57	Cases	20	12
	Controls	61	183
57–75	Cases	37	45
	Controls	113	378
> 75	Cases	9	42
	Controls	23	140

Use the Mantel-Haenszel procedure to compare the cases versus the controls.

15. Data taken from a study to investigate the effects of smoking on cervical cancer are stratified by the number of sexual partners. Results are as follows:

No. of partners	Smoking	Cancer	
		Yes	No
Zero or one	Yes	12	21
	No	25	118
Two or more	Yes	96	142
	No	92	150

Use the Mantel-Haenszel procedure to compare the cases versus controls.

16. The criterion for issuing a smog alert is established at greater than 7 ppm of a particular pollutant. Samples collected from 16 stations in a certain city give an $\bar{x}$ of 7.84 ppm with a standard deviation of $s = 2.01$ ppm. Do these findings indicate that the smog alert criterion has been exceeded?

17. The purpose of an experiment is to investigate the effect of vagal nerve stimulation on insulin secretion. The subjects are mongrel dogs with varying body weights. The following table gives the amount of immunoreactive insulin in pancreatic venous plasma just before stimulation of the left vagus and the amount measured 5 minutes after stimulation for 7 dogs.

Dog	Blood levels of immunoreactive insulin (μU/ml)	
	Before	After
1	350	480
2	200	130
3	240	250
4	290	310
5	90	280
6	370	1,450
7	240	280

Test the null hypothesis that the stimulation of the vagus nerve has no effect on the blood level of immunoreactive insulin, i.e.,

$$\mathcal{H}_0 : \mu_{\text{before}} = \mu_{\text{after}}$$

Use $\alpha = .05$.

18. A study was conducted to investigate whether oat bran cereal helps to lower serum cholesterol in men with high cholesterol levels. Fourteen men were randomly placed on a diet that included either oat bran or corn flakes; after 2 weeks, their low-density lipoprotein cholesterol levels were recorded. The data were:

Subject	LDL (mmol/liter)	
	Corn flakes	Oat bran
1	4.61	3.84
2	6.42	5.57
3	5.40	5.85
4	4.54	4.80
5	3.98	3.68
6	3.82	2.96
7	5.01	4.41
8	4.34	3.72
9	3.80	3.49
10	4.56	3.84
11	5.35	5.26
12	3.89	3.73
13	2.25	1.84
14	4.24	4.14

State and test your null hypothesis using the two-sample t-test.

19. The Australian study of Example 6.13 also provided these data on monocular acuity (expressed in log scale) for two female groups of subjects:

• Australian females of European origin

$$n_1 = 63$$

$$\bar{x}_1 = -.13$$

$$s_1 = .17$$

• Australian females of Aboriginal origin

$$n_2 = 54$$

$$\bar{x}_2 = -.24$$

$$s_2 = .18$$

Do these indicate a racial variation among women?

20. In a trial to compare a stannous fluoride dentifrice (A) with a commercially available fluoride-free dentifrice (D), 270 children received A and 250 received D for a period of 3 years. The number x of DMFS increments (that is, the number of new Decayed, Missing, and Filled Tooth Surfaces) was obtained for each child. Results were:

$$\text{Dentifrice A:} \quad \bar{x}_A = 9.78$$

$$s_A = 7.51$$

$$\text{Dentifrice D:} \quad \bar{x}_D = 12.83$$

$$s_D = 8.31$$

Do the results provide strong enough evidence to suggest a real effect of fluoride in *reducing* the mean DMFS?

21. An experiment was conducted at the University of California at Berkeley to study the psychological environment effect on the anatomy of the brain. A group of 19 rats was randomly divided into two groups. Twelve animals in the treatment group lived together in a large cage, furnished with playthings that were changed daily, while animals in the control group lived in isolation with no toys. After a month, the experimental animals were killed and dissected. The following table gives the cortex weights (the thinking part of the brain) in milligrams:

Treatment	Control
707	669
740	650
745	651
652	627
649	656
676	642
699	698
696	
712	
708	
749	
690	

Use the two-sample t-test to compare the means of the two groups.

22. Depression is one of the most commonly diagnosed conditions among hospitalized patients in mental institutions. The occurrence of depression was determined during the summer of 1979 in a multiethnic probability sample of 1,000

adults in Los Angeles County, as part of a community survey of the epidemiology of depression and help-seeking behavior. The primary measure of depression was the CES-D scale developed by the Center for Epidemiologic Studies. On a scale of 0 to 60, a score of 16 or higher was classified as depression. The following table gives the average CES-D score for the two sexes.

	CES-D score		
	Cases	$\bar{x}$	s
Male	412	7.6	7.5
Female	588	10.4	10.3

Use a t-test to compare the males versus the females.

23. Data in epidemiologic studies are sometimes self-reported. Screening data from the hypertension detection and follow-up program in Minneapolis, Minnesota (1973–1974) provided an opportunity to evaluate the accuracy of self-reported height and weight. The following table gives the percent discrepancy between self-reported and measured height:

$$x = \frac{\text{Self-reported height} - \text{measured height}}{\text{Measured height}} \times 100\%$$

	Men			Women		
Education	n	mean	SD	n	mean	SD
$\leq$ High school	476	1.38	1.53	323	.66	1.53
$\geq$ College	192	1.04	1.31	62	.41	1.46

Compare the mean difference in percent discrepancy between

(a) Men with different education levels

(b) Women with different education levels

(c) Men versus women at each educational level

24. A case–control study was undertaken to study the relationship between hypertension and obesity. Persons aged 30–49 years who were clearly non-hypertensive at their first multiphasic health checkup and became hypertensive by age 55 years were sought and identified as cases. Controls were selected from among participants in a health plan, those who had the first checkup and no sign of hypertension in subsequent checkups. One control was matched to each case

based on sex, race, year of birth, and year of entrance into the health plan. Data for 609 matched pairs are summarized as follows.

Variable	Paired difference	
	Mean	Standard deviation
Systolic blood pressure (mmHg)	6.8	13.86
Diastolic blood pressure (mmHg)	5.4	12.17
Body mass index (kg/m^2)	1.3	4.78

Compare the cases versus the controls using each measured characteristic.

25. A study was undertaken regarding the relationship between exposure to polychlorinated biphenyls (PCBs) and reproduction among women occupationally exposed to PCBs during the manufacture of capacitors in up-state New York. Interviews were conducted in 1982 with women who had held jobs with direct exposure and women who had never held a direct-exposure job to ascertain information on reproductive outcomes. Data are summarized in the following table.

Variable	Exposure			
	Direct ($n = 172$)		Indirect ($n = 184$)	
	Mean	SD	Mean	SD
Weight gain during pregnancy (lbs)	25.5	14.0	29.0	14.7
Birth weight (g)	3,313	456	3,417	486
Gestational age (days)	279.0	17.0	279.3	13.5

Test to evaluate the effect of direct exposure (as compared with indirect exposure) using each measured characteristic.

26. A study was done to determine if simplification of smoking literature improved patient comprehension. All subjects were administered a pretest. Subjects were then randomized into three groups. One group received no booklet, one group received one written at the 5th grade reading level, and the third received one written at the 10th grade reading level. After booklets were received, all subjects were administered a second test. The mean score differences (postscore−prescore) are given below along with their standard deviations and sample sizes.

	No booklet	5th grade level	10th grade level
$\bar{x}$	0.25	1.57	0.63
s	2.28	2.54	2.38
n	44	44	41

Compare the groups with booklet to the group with no booklet.

27. A study was conducted to investigate the risk factors for peripheral arterial disease among persons 55–74 years of age. The following table provides data on LDL cholesterol levels (mmol/liter) from four different subgroups of subjects:

Group	n	$\bar{x}$	s
1. Patients with intermittent claudication	73	6.22	1.62
2. Major asymptotic disease cases	105	5.81	1.43
3. Minor asymptotic disease cases	240	5.77	1.24
4. Those with no disease	1,080	5.47	1.31

Compare each disease group (groups 1, 2, and 3) versus the group with no disease.

28. The following data are the rates of oxygen consumption of eight birds measured at different environmental temperatures.

x, temperature (°C)	y, oxygen consumption (ml/g/hr)
−19	5.2
−15	4.7
−10	4.5
−5	3.6
0	3.4
5	3.1
10	2.7
19	1.8

Prepare a scatter diagram and calculate the coefficient of correlation r.

29. In a study of different regimens in the management of diabetes, one outcome of concern was weight loss during the course of therapy. The following data pertain to the question of whether the amount of weight loss is related to ini-

tial weight. For 16 newly diagnosed adult diabetic patients who received phenformins to manage their diabetic state, their initial weight at the start of therapy (denoted by x lbs) and their weight loss 1 year after therapy began (denoted by y lbs—a weight gain is expressed as a negative weight "loss") were obtained and summarized into

$$\sum(x - \bar{x})^2 = 20{,}192$$

$$\sum(y - \bar{y})^2 = 3{,}380$$

$$\sum(x - \bar{x})(y - \bar{y}) = 5{,}786$$

Test for the null hypothesis of independence at $\alpha = .05$.

30. Nutrition studies consistently show positive associations between dietary fats and serum cholesterol levels. However, this relationship is hard to assess accurately, partly due to the highly variable habits of individuals. A study was conducted to examine this relationship among a group of vegetarians.

Saturated and polyunsaturated fatty acids and dietary cholesterol were combined into a "Keys dietary score" (X). Using serum cholesterol (mg/dl) as the dependent variable (Y), data were summarized as follows:

$$n = 46$$

$$r = .46$$

Test for the null hypothesis of independence.

CHAPTER

7

Introduction to Other Selected Topics

7.1. MORE ON EXPLANATORY DATA ANALYSIS

This section introduces a few more methods on explanatory data analysis as supplemental to the graphical methods of Chapter 2. These topics are not more advanced; they are only newer, and a few have gained popularity only in recent years. Some of these methods, for example, the one-way scatter plots and the box plots, require access to computers for easy implementation.

7.1.1. One-Way Scatter Plots

The one-way scatter plot is another type of graph that can be used to summarize a set of continuous observations. A one-way scatter plot uses a single horizontal axis to display the relative position of each data point. As an example, Figure 7.1 depicts the crude death rates for all 50 states and the District of Columbia, from a low of 393.9 per 100,000 population to a high of 1,242.1 per 100,000 population.

Fig. 7.1   Crude death rates for the United States, 1988.

An advantage of a one-way scatter plot is that, since each observation is represented individually, no information is lost; a disadvantage is that it may be difficult to read (and to construct!) if values are close together.

7.1.2. Box Plots

The box plot is a graphical representation of a data set that gives a visual impression of location, spread, as well as the degree and direction of skewness. It also allows for the identification of outliers. Box plots are similar to one-way scatter plots in that they require a single horizontal axis; however, instead of plotting each and every observation they display a summary of the data. A box plot consists of the following:

1. A central box extends from the 25th to the 75th percentiles. This box is divided into two compartments at the median value of the data set. The relative sizes of the two halves of the box provide an indication of the distribution symmetry. If they are approximately equal, the data set is roughly symmetric; otherwise, we are able to see the degree and direction of skewness (Figure 7.2).
2. The line segments projecting out from the box extend in both directions to the so-called *adjacent values*. The adjacent values are the points that are 1.5 times the length of the box beyond either quartile. All other data points outside this range are represented individually by little circles; these are considered to be outliers or extreme observations that are not typical of the rest of the data.

Of course, it is possible to combine a one-way scatter plot and a box plot so as to convey an even greater amount of information (Figure 7.3). There are other ways

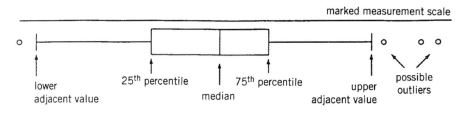

Fig. 7.2 A typical box plot.

Fig. 7.3 Crude death rates for the United States, 1988. A combination of one-way scatter and box plots.

TABLE 7.1 Weights (lbs) of 57 Children in Day Care

68	63	42	27	30	36	28	32	79	27
22	23	24	25	44	65	43	25	74	51
36	42	28	31	28	25	45	12	57	51
12	32	49	38	42	27	31	50	38	21
16	24	69	47	23	22	43	27	49	28
23	19	46	30	43	49	12			

of constructing box plots; for example, one may make it vertically or divided into different levels of outliers.

7.1.3. Stem-and-Leaf Diagrams

A stem-and-leaf diagram is a graphical representation in which the data points are grouped in such a way that we can see the shape of the distribution while retaining the individual values of the data points. Stem-and-leaf diagrams are similar to the frequency tables and histograms of Chapter 2, but they also display each and every observation. Data on the weights of children from Example 2.2 are adopted here to illustrate the construction of this simple device. The weights (in lbs) of 57 children at a day care center are given in Table 7.1.

A stem-and-leaf diagram consists of a series of rows of numbers. The number used to label a row is called a *stem,* and the other numbers in the row are called *leaves*. There are no hard rules about how to construct a stem-and-leaf diagram. Generally, it consists of the following steps:

1. Choose some convenient/conventional numbers to serve as stems. The stems chosen are usually the first one or two digits of individual data points.
2. Reproduce the data graphically by recording the digit or digits following the stems as a leaf on the appropriate stem.

If the final graph is turned on its side, it looks similar to a histogram (Figure 7.4).

```
1 |              1 | 2 2 2 6 9
2 |              2 | 1 2 2 3 3 3 4 4 5 5 5 7 7 7 7 8 8 8 8
3 |              3 | 0 0 1 1 2 2 6 6 8 8
4 |      ⇒       4 | 2 2 2 3 3 3 4 5 6 7 9 9 9
5 |              5 | 0 0 1 7
6 |              6 | 3 5 8 9
7 |              7 | 4 9
Stems        Stems   Leaves
```

Fig. 7.4 A typical stem-and-leaf diagram.

7.2. SAMPLE SIZE DETERMINATION

The determination of the size of a sample is a crucial element in the design of a survey or a clinical trial. In designing any study, one of the first questions that must be answered is, "How large must the sample be to accomplish the goals of the study?" Depending on the study goals, the planning of sample size can be approached in two different ways: either in terms of controlling the width of a desired confidence interval for the parameter of interest or in terms of controlling the risk of making Type II errors.

7.2.1. Estimation of Means and Proportions

For the confidence interval to be useful, it must be short enough to pinpoint the value of the parameter, such as the mean μ or the population proportion π, reasonably well with a high degree of confidence. If a study is unplanned or poorly planned, there is a real possibility that the resulting confidence interval will be too long to be of any use to the researcher. For example, a study that allows us to conclude that "we are 95% confident that the true average birth weight of a baby boy lies between 2 and 20 pounds" is clearly not of much value. Half the width of a confidence interval is often called the *error* of the estimate. The shorter the confidence interval, the smaller the error, the better the estimate, and, the larger the sample size we need.

Estimation of a Mean

Suppose the goal of a study is to estimate an unknown population mean μ; say, the average birth weight of babies born to mothers addicted to cocaine. How large must the sample be? Of course, the answer depends on how accurate we want our estimate to be. Let us suppose we wish to have the resulting 95% confidence interval for μ no longer than $2d$ units of measurement. In other words, we want the "error" of our estimate less than or equal to d. For example, in estimating the average birth weight of babies born to mothers addicted to cocaine, we want to be correct within, say, half a pound.

Recall from Chapter 4 that the 95% confidence interval for a mean μ is

$$\bar{x} \pm 1.96 \frac{s}{\sqrt{n}}$$

(A different coefficient is used if the degree of confidence is not 95%. For example, if 99% confidence is desired, multiply the standard error of the mean $s/\sqrt{n}$ by 2.58.) Therefore, the error of the estimate is

$$1.96 \frac{s}{\sqrt{n}}$$

It is now seen that the larger the sample size n, the smaller the error. Our goal is to make

$$1.96 \frac{s}{\sqrt{n}} \leq d$$

The minimum sample size needed is

$$n = \frac{(1.96)^2 s^2}{d^2}$$

(rounded up to the next integer). This required sample size is affected by three things:

(i) The coefficient 1.96. (As previously mentioned, a different coefficient is used for a different degree of confidence, which is set arbitrarily by the investigator; 95% is a conventional choice.)

(ii) The maximum "tolerated error" d, which is also set arbitrarily by the investigator.

(iii) The variability of the population measurements, the variance.

This seems like a circular problem. We want to find a sample so as to estimate the mean accurately, and in order to do that, we need to know the variance! Of course, the exact value of the variance is also unknown. However, we can use information from similar studies, past studies, or some reasonable "upper bound." If nothing else is available, we may need to run a preliminary or pilot study. One-fourth of the range may serve as a rough estimate for the standard deviation.

_____

Example 7.1

Suppose that a study is to be conducted to estimate the average birth weight of babies born to mothers addicted to cocaine. Suppose also that we want to estimate this average to within a half pound with 95% confidence. This goal specifies two quantities:

$$d = .5$$

$$\text{Coefficient} = 1.96$$

What value should be used for the variance? Let us try two different approaches.

(i) It is probably safe to assume that the birth weight of these babies lies between 2 and 12 lbs. A rough estimate of s is

$$s \cong \frac{12 - 2}{4}$$

$$= 2.5 \text{ lb}$$

(ii) Information from normal babies may be used to estimate s. The rationale here is that the addiction affects every baby almost uniformly; this may result in a smaller average but the variance is unchanged.

If we use

$$s \cong 2.5$$

then the required sample size is

$$n = \frac{(1.96)^2 (2.5)^2}{(.5)^2}$$
$$= 96.04$$

Since we can't sample 96.04 babies, we will take a sample of size 97. (Note that if it turns out that $s \cong 2$ lbs, then only 62 babies are needed. It is not an exact science; we just do the best we can without adequate information about the population variance.)

Estimation of a Proportion

Suppose the goal of another study is to estimate an unknown population proportion π, say, the smoking rate of a certain well-defined population. In this case, we also want to have an error of the estimate not exceeding d.

Recall, also from Chapter 4, that the 95% confidence interval for the population proportion π is

$$p \pm 1.96 \sqrt{\frac{p(1-p)}{n}}$$

where p is the sample proportion. Therefore, our goal is expressed as

$$1.96 \sqrt{\frac{p(1-p)}{n}} \leq d$$

leading to the required minimum sample size

$$n = \frac{(1.96)^2 p(1-p)}{d^2}$$

(rounded up to the next integer). This required sample size is also affected by three factors:

(i) The coefficient, 1.96
(ii) The maximum tolerated error, d
(iii) The proportion p itself

This third factor is now more unsettling. To find n so as to obtain an accurate value of the proportion, we need the proportion itself. There is no perfect, exact solution for this. Usually, we can use information from similar studies, past studies, or studies on similar populations. If no good prior knowledge about the proportion is available, we can replace $p(1 - p)$ by .25 and use a "conservative" sample size

$$n_c = \frac{(1.96)^2(.25)}{d^2}$$

because $n_c \geq n$ regardless of the value of π.

Example 7.2

Suppose that a study is to be conducted to estimate the smoking rate among NOW (National Organization for Women) members. Suppose also that we want to estimate this proportion to within 3% (i.e., $d = .03$) with 95% confidence.

(i) Since the current smoking rate among women in general is about 27% or .27, we can use this figure in calculating the required sample size. This results in

$$n = \frac{(1.96)^2(.27)(.73)}{(.03)^2}$$
$$= 841.3$$

or a sample of size 842 is needed.

(ii) If we do not want or have the above figure of 27%, then we still can conservatively take

$$n_c = \frac{(1.96)^2(.25)}{(.03)^2}$$
$$= 1,067.1$$

i.e., we can sample 1,068 members of NOW. Note that this "conservative" sample size is adequate regardless of the true value π of the unknown population proportion; values of n and n_c are closer when π is near .5.

7.2.2. Comparison of Two Means

In the previous section, the planning of sample size was approached in terms of controlling the width of a desired confidence interval for the parameter of interest, either the population mean or proportion. However, clinical trials are often conducted not for parameter estimation, but for the comparison of two treatments, e.g., a new therapy versus a placebo or a standard therapy. Therefore, it is more suitable to approach the planning of sample size in terms of controlling the risk of making Type II errors.

For example, a researcher is studying a drug that is to be used to reduce the cholesterol level in adult males aged 30 years and over. Subjects are to be randomized into two groups, one receiving the new drug (group 1) and one a look-alike placebo (group 2). The response variable considered is the change in cholesterol level before and after the intervention. The null hypothesis to be tested is

$$\mathcal{H}_0 : \mu_1 = \mu_2$$

versus

$$\mathcal{H}_A : \mu_2 > \mu_1$$

How large a total sample should be used to conduct the study? (In a balanced randomization, each group consists of $n/2$ subjects.)

Recall that in testing a null hypothesis, two types of errors are possible. We might reject $\mathcal{H}_0$ when in fact $\mathcal{H}_0$ is true, thus committing a Type I error. However, this type of error can be controlled in the decision-making process; conventionally, the probability of making this mistake is set at $\alpha = .05$ or $\alpha = .01$. A Type II error occurs when we fail to reject $\mathcal{H}_0$ even though it is false. In the above drug testing example, a Type II error leads to our inability to recognize the effectiveness of the new drug being studied. The probability of committing a Type II error is denoted by β, and $(1 - \beta)$ is called the *power* of a statistical test. Since the power is the probability that we will be able to support our research claim (i.e., the alternative hypothesis) when it is right, studies should be designed to have high power. This is achieved through the planning of sample size.

In the comparison of two population means, μ_1 versus μ_2, the required minimum total sample size is calculated from

$$N = 4(z_{1-\alpha} + z_{1-\beta})^2 \frac{\sigma^2}{d^2}$$

assuming that we conduct a balanced study with each group consisting of $n = N/2$ subjects. This required total sample size is affected by four factors:

(i) The "size" α of the test. As previously mentioned, this is set arbitrarily by the investigator; conventionally, $\alpha = .05$ is often used. The quantity $z_{1-\alpha}$ in the above formula is the percentile of the standard normal distribution associated with a choice of α; for example, $z_{1-\alpha} = 1.96$ when $\alpha = .05$ is chosen.

(ii) The desired power $(1 - \beta)$ (or probability of committing a Type II error β). This value is also selected by the investigator; a power of 80% or 90% is often used.

(iii) The quantity

$$d = |\mu_2 - \mu_1|$$

which is the magnitude of the difference between μ_1 and μ_2 that is deemed to be important.

(iv) The variance σ^2 of the population. Similar to the problem of parameter estimation, this variance is the only quantity that is difficult to determine. Again, the exact value of σ^2 is unknown; we may use information from similar studies or past studies. Sometimes, we may even need to run a preliminary or pilot study to estimate σ^2.

Example 7.3

Suppose a researcher is studying a drug that is used to reduce the cholesterol level in adult males aged 30 years or over and wants to test it against a placebo in a balanced randomized study. Suppose also that it is important that a reduction difference of 5 be detected ($d = 5$). We decide to preset $\alpha = .05$ and want to design a study such that its power to detect a difference between means of 10 is 90% (or $\beta = .10$). Also, the variance of cholesterol reduction (with placebo) is known to be about $s^2 \cong 36$. From Appendix B,

$$\alpha = .05 \rightarrow z_{1-\alpha} = 1.96$$

$$\beta = .10 \rightarrow z_{1-\beta} = 1.65$$

leading to the required total sample size

$$N = 4(1.96 + 1.65)^2 \frac{36}{25}$$

$$\cong 76$$

Each group will have 38 subjects.

7.2.3. Comparison of Two Proportions

Let us consider a similar problem where we want to design a study to compare two proportions. For example, a new vaccine will be tested in which subjects are to be randomized into two groups of equal size: a control (unimmunized) group (group 1) and an experimental (immunized) group (group 2). Subjects, in both control and experimental groups, will be challenged by a certain type of bacteria, and we wish to compare the infection rates. The null hypothesis to be tested is

$$\mathcal{H}_0 : \pi_1 = \pi_2$$

versus

$$\mathcal{H}_A : \pi_1 < \pi_2$$

How large a total sample should be used to conduct this vaccine study?
Suppose that it is important to detect a reduction of infection rate

$$d = \pi_2 - \pi_1$$

If we decide to pre-set the size of the study at $\alpha = .05$ and want the power $(1 - \beta)$ to detect the difference d, then the required sample size is given by this complicated formula:

$$N = \frac{4\{2z_{1-\alpha}\pi(1-\pi) + z_{1-\beta}[\pi_1(1-\pi_1) + \pi_2(1-\pi_2)]\}^2}{(\pi_2 - \pi_1)^2}$$

In this formula the quantities $z_{1-\alpha}$ and $z_{1-\beta}$ are defined as in the previous section; π is the common value of the proportions under $\mathcal{H}_0$. It is obvious that the problem of planning sample size is more difficult, and a good solution requires a deeper knowledge of the scientific problem: some good idea of the magnitude of the proportions π_1 and π_2 themselves.

Example 7.4

A new vaccine will be tested in which subjects are to be randomized into two groups of equal size: a control (unimmunized) group and an experimental (immunized) group. Based on prior knowledge about the vaccine through small pilot studies, the following assumptions are made:

(i) The infection of the control group (when challenged by a certain type of bacteria) is expected to be about 50%, i.e.,

$$\pi_2 = .50$$

(ii) About 80% of the experimental group is expected to develop adequate antibodies (that is, at least a twofold increase). If antibodies are inadequate, then the infection rate is about the same as for a control subject. But if an experimental subject has adequate antibodies, then the vaccine is expected to be about 85% effective (that corresponds to a 15% infection rate against the challenged bacteria).

Putting these assumptions together, we obtain an expected value of π_1:

$$\pi_1 = (.80)(.15) + (.20)(.50)$$

$$= .22$$

Suppose also that we decide to pre-set $\alpha = .05$ and want the power to be about 90% (i.e., $\beta = .10$). In other words, we use

$$z_{1-\alpha} = 1.96$$

$$z_{1-\beta} = 1.65$$

From this information, the required total sample size is

$$N = \frac{4\{(2)(1.96)(.50)(.50) + (1.65)[(.50)(.50) + (.22)(.78)]\}^2}{(.50 - .22)^2}$$

$$\cong 144$$

so that each group will have 72 subjects. In this solution we use

$$\pi = .50$$

because, under the null hypothesis, the vaccine is not effective (or is only as effective as the control) so that the common value of the infection rate is that of the control group, 50%.

7.3. NONPARAMETRIC METHODS

In the previous chapters and sections, we have seen how to apply statistical procedures that make the most efficient use of the data from a study. However, these methods depended on certain assumptions about the distributions in the population; for example, a certain procedure may be derived assuming that the population is normally distributed. To a certain degree, these procedures are robust; that is, they are relatively insensitive to departures from the assumptions made. In other words, departures from those assumptions have only very little effect on the results, provided that samples are large enough. But the procedures are all sensitive to extreme observations, a few very small or very large—perhaps erroneous—data values. In this very last section of the book, we will learn some nonparametric procedures, or distribution-free methods, where no assumptions about population distributions are made.

7.3.1. The Wilcoxon Rank-Sum Test

The Wilcoxon rank-sum test is perhaps the most popular nonparametric procedure. The Wilcoxon test is a nonparametric counterpart of the two-sample t-test; it is used to compare two samples that have been drawn from independent populations. But, unlike the t-test, the Wilcoxon test does not assume that the underlying populations are normally distributed and is less affected by extreme observations. The Wilcoxon rank sum test evaluates the null hypothesis that the medians of the two populations are identical (for a normally distributed population, the population median is also the population mean).

For example, a study was designed to test the question of whether cigarette smoking is associated with reduced serum testosterone levels. To carry out this research objective two samples, each of size 10, are independently selected. The first sample consists of 10 non-smokers who have never smoked and the second sample 10 heavy smokers, defined as those who smoke 30 or more cigarettes per day. To perform the Wilcoxon rank sum test, we combine the two samples into one large sample (of size 20), arrange the observations from smallest to largest, and assign a rank, from 1 to 20, to each one. If there are tied observations, then we assign an average rank to all measurements with the same value. For example, if the two observations next to the third smallest are equal, we assign an average rank of $(4 + 5)/4.5$ to each one. The next step is to find the sum of the ranks corresponding to each of the original samples. Let n_1, n_2 be the two sample sizes and R be the sum of the ranks from the sample with size n_1.

Under the null hypothesis that the two underlying populations have identical medians, we would expect the averages of ranks to be approximately equal. We test this hypothesis by calculating the statistic

$$z = \frac{R - \mu_R}{\sigma_R}$$

where

$$\mu_R = \frac{n_1(n_1 + n_2 + 1)}{2}$$

is the mean and

$$\sigma_R = \sqrt{\frac{n_1 n_2(n_1 + n_2 + 1)}{12}}$$

is the standard deviation of R. It does not make any difference which rank sum we use. For relatively large values of n_1 and n_2 (say, both greater than or equal to 10), the sampling distribution of this statistic is approximately standard normal. The null hypothesis is rejected at the 5% level if

$$z < -1.96 \quad \text{or} \quad z > 1.96$$

Example 7.5

For the above study on cigarette smoking, the following table shows the raw data, where testosterone levels were measured in μg/dl and the ranks were as follows:

Non-smokers		Heavy smokers	
Measurement	Rank	Measurement	Rank
.44	8.5	.45	10
.44	8.5	.25	1
.43	7	.40	6
.56	14	.27	2
.85	17	.34	4
.68	15	.62	13
.96	20	.47	11
.72	16	.30	3
.92	19	.35	5
.87	18	.54	12

The sum of the ranks for group 1 (non-smokers) is

$$R = 143$$

In addition,

$$\mu_R = \frac{10(10 + 10 + 1)}{2}$$
$$= 105$$

and

$$\sigma_R = \sqrt{\frac{(10)(10)(10 + 10 + 1)}{12}}$$
$$= 13.23$$

Substituting these values into the equation for the test statistic, we have

$$z = \frac{R - \mu_R}{\sigma_R}$$
$$= \frac{143 - 105}{13.23}$$
$$= 2.87$$

Since $z > 1.96$, we reject the null hypothesis at the 5% level. (In fact, since $z > 2.58$, we reject the null hypothesis at the 1% level.) Note that if we use the sum of the ranks for the other group (heavy smokers; the sum of the ranks is 67) we have

$$\frac{67 - 105}{13.23} = -2.87$$

and we would come to the same decision.

7.3.2. Ordered $2 \times k$ Contingency Tables

This section presents an efficient method for use with ordered $2 \times k$ contingency tables, tables with 2 rows and with k columns having a certain natural ordering. Let us first consider an example concerning the use of seat belts in automobiles. Each accident in this example is classified according to whether a seat belt was used and the severity of injuries received: none, minor, major, or death. The data are given in Table 7.2. To compare the extent of injuries of those who used seat belts and those who did not, we can perform a chi-square test as explained in Chapter 6. Such an application of the chi-square test yields

$$\chi^2 = 9.26$$

whereas the cut-point for the chi-square with 3 degrees of freedom is 7.81 (for $\alpha = .05$). Therefore, the difference between the two groups is significant at the 5% level. But the usual chi-square calculation takes no account of the fact that the extent of injury has a natural ordering: none < minor < major < death. In addition, the percent of seat belt users in each injury group decreases from level "none" to level "death":

$$\text{None}: \quad 75/(75 + 65) = 58\%$$

$$\text{Minor}: \quad 160/(100 + 175) = 48\%$$

$$\text{Major}: \quad 100/(100 + 135) = 43\%$$

$$\text{Death}: \quad 15/(15 + 25) = 38\%$$

TABLE 7.2 Possible Association Between Injury and Use of Seat Belt

| Seat belt | Extent of injury received | | | |
	None	Minor	Major	Death
Yes	75	160	100	15
No	65	175	135	25

TABLE 7.3   Typical $2 \times k$ Table

Row	1	2	$\cdots$	k	Total
		Column level			
1	a_1	a_2	$\cdots$	a_k	A
2	b_1	b_2	$\cdots$	b_k	B
Total	n_1	n_2	$\cdots$	n_k	N

We now present a special procedure specifically designed to detect such a "trend" and will use the same example to show that it attains a higher degree of significance. In general, consider an ordered $2 \times k$ table with the frequencies shown in Table 7.3. The number of "concordances" is calculated by

$$C = a_1(b_2 + \cdots + b_k) + a_2(b_3 + \cdots + b_k) + \cdots + a_{k-1}b_k$$

(The term "concordance" pair as used in the above example corresponds to a less severe injury for the seat belt user.) The number of "discordances" is

$$D = b_1(a_2 + \cdots + a_k) + b_2(a_3 + \cdots + a_k) + \cdots + b_{k-1}a_k$$

In order to perform the test, we calculate the statistic

$$S = C - D$$

then standardize it to obtain

$$z = \frac{S - \mu_S}{\sigma_D}$$

where $\mu_S = 0$ is the mean of S under the null hypothesis and

$$\sigma_S = \left[\frac{AB}{3N(N-1)} (N^3 - n_1^3 - n_2^3 - \cdots - n_k^3) \right]^{1/2}$$

The standardized z-score is distributed as standard normal if the null hypothesis is true. The null hypothesis is rejected at the 5% level if

$$z \leq -1.96 \quad \text{or} \quad z \geq 1.96$$

Example 7.6

For the above study on the use of seat belts in automobiles, we have

$$C = 75(175 + 135 + 25) + 160(135 + 25) + (100)(25)$$

$$= 53,225$$

$$D = 65(160 + 100 + 15) + 175(100 + 15) + (135)(15)$$

$$= 40,025$$

In addition, we have

$$A = 350$$

$$B = 390$$

$$n_1 = 130$$

$$n_2 = 335$$

$$n_3 = 235$$

$$n_4 = 40$$

$$N = 740$$

Substituting these values into the equations of the test statistic, we have

$$S = 53,225 - 40,025$$

$$= 13,200$$

$$\sigma_S = \left[\frac{(350)(390)}{(3)(740)(739)} (740^3 - 130^3 - 335^3 - 235^3 - 40^3) \right]^{1/2}$$

$$= 5,414.76$$

leading to

$$z = 13{,}200/5{,}414.76$$

$$= 2.44$$

which shows a higher degree of significance ($z > 1.96$) than the chi-square test.

_____

7.3.3. Rank Correlations

Suppose the data set consists of n pairs of observations $[(x_i, y_i)]$ expressing a possible relationship between two continuous variables. We characterize the strength of such a relationship by calculating the coefficient of correlation

$$r = \frac{\sum(x - \bar{x})(y - \bar{y})}{\sqrt{[\sum(x - \bar{x})^2][\sum(y - \bar{y})^2]}}$$

(called the *Pearson's correlation coefficient*), as shown in Chapter 6. Like other "parametric" techniques—the t-test, the mean—the correlation coefficient r is very sensitive to extreme observations. We may be interested in calculating a measure of association that is more robust with respect to outlying values. There are not one but two nonparametric procedures: the *Spearman's rho* and the *Kendall's tau* rank correlations.

The Spearman's Rho

The Spearman's rank correlation is a direct nonparametric counterpart of the Pearson's correlation coefficient. To perform this procedure, we first arrange the x values from smallest to largest and assign a rank from 1 to n for each value; let R_i be the rank of value x_i. Similarly, we arrange the y values also from smallest to largest and assign a rank from 1 to n for each value; let S_i be the rank of value y_i. If there are tied observations, we assign an average rank just as in the formation of the Wilcoxon rank sum test. The next step is to replace, in the formula of the Pearson's correlation coefficient r, x_i by its rank R_i and y_i by its rank S_i. The result is the Spearman's rho, a popular rank correlation:

$$\rho = \frac{\sum(R_i - \bar{R})(S_i - \bar{S})}{\sqrt{[\sum(R_i - \bar{R})^2][\sum(S_i - \bar{S})^2]}}$$

$$= 1 - \frac{6\sum(R_i - S_i)^2}{n(n^2 - 1)}$$

The second expression is simpler and easier to use.

Example 7.7

Consider again the birth weight problem of Example 6.16. We have

Birth weight		Increase in weight			
x (oz)	Rank R	y (%)	Rank S	R − S	$(R − S)^2$
112	10	63	3	7	49
111	9	66	4	5	25
107	8	72	5.5	2.5	6.25
119	12	52	2	10	100
92	4	75	7	−3	9
80	1	118	11	−10	100
81	2	120	12	−10	100
84	3	114	10	−7	49
118	11	42	1	10	100
106	7	72	5.5	1.5	2.25
103	6	90	8	−2	4
94	5	91	9	−416	
Total					560.50

Substituting the value of $\sum(R_i - S_i)^2$ into the formula for rho (ρ), we obtain

$$\rho = 1 - \frac{(6)(560.5)}{(12)(143)}$$

$$= -.96$$

which is very close to the value of r (−.946) obtained in Chapter 6 (Example 6.16). This closeness is true when there are few or no extreme observations.

The Kendall's tau

Unlike the Spearman's rho, the other rank correlation—the Kendall's tau τ—is defined and calculated very differently, even though they often yield similar numerical results. The above birth weight problem (Examples 6.16 and 7.7) is adapted to illustrate this method with the following steps:

1. First, the x- and y-values are presented in two rows; the x-values in the first row are arranged from smallest to largest.

2. For each y-value in the second row, we count

   (a) The number of larger y-values to its right (third row). The sum of these is denoted by C.
   (b) The number of smaller y-values to its right (fourth row). The sum of these is denoted by D.

   C and D are the numbers of concordant pairs and discordant pairs, respectively; the basic idea is similar to that used for the ordered $2 \times k$ tables.
3. The Kendall's rank correlation is defined by

$$\tau = \frac{C - D}{\frac{1}{2}n(n - 1)}$$

Example 7.8

For the above birth weight problem, we have

													Total
x	80	81	84	92	94	103	106	107	111	112	118	119	
y	118	120	114	75	91	90	72	72	66	63	42	52	
C	1	0	0	2	0	0	0	0	0	0	0	0	3
D	10	10	9	6	7	6	4	4	3	2	0	0	61

The value of Kendall's tau is

$$\tau = \frac{3 - 61}{\frac{1}{2}(12)(11)}$$

$$= -.88$$

EXERCISES

1. Normal red blood cells in humans are shaped like biconcave disks. Occasionally, hemoglobin, a protein that readily combines with oxygen, is imperfectly formed in the cell. One type of imperfect hemoglobin causes the cells to have a caved-in, or "sickle-like" appearance. These "sickle" cells are less efficient carriers of oxygen than normal cells and result in an oxygen deficiency called *sickle cell anemia*. This condition has a significant prevalence among blacks. Suppose a study is to be conducted to estimate this prevalence among blacks in a certain large city.

(a) How large a sample should be chosen to estimate this proportion to within 1 percentage point with 99% confidence? With 95% confidence? (No prior estimate of the prevalence in this city is assumed available.)

(b) A similar study was recently conducted in another state. Of the 13,573 blacks sampled, 1,085 were found to have sickle cell anemia. Using this information, re-solve problem (a).

2. How many patients must be sampled to estimate the proportion of children with chest pain who had a normal chest x-ray? No prior knowledge is assumed available.

3. How large a sample is needed to estimate the average total protein level among adults to within .5 g/dl with 95% confidence if these values are known to have a range of approximately 2.5 g/dl?

4. A researcher wants to estimate the average weight loss obtained by patients at a residential weight loss clinic during the first week of a controlled diet and exercise regimen. How large a sample is needed to estimate this mean to within .5 lb with 95% confidence? Assume that past data indicate a standard deviation of about 1 lb.

5. Suppose we want to estimate the average caloric intake by college students to within 100 calories with 95% confidence. If these values are known to range from 500 to 4,000 calories, how large a sample is needed to make this estimate?

6. A study will be conducted to investigate a claim that oat bran will reduce serum cholesterol in men with high cholesterol levels. Subjects will be randomized to diets that include either oat bran or corn flakes cereals. After 2 weeks, LDL cholesterol level (in mmol/liter) will be measured and the two groups will be compared via a two-sample t-test. A pilot study ($n = 14$) with corn flakes yields

$$\bar{x} = 4.44, \qquad s = .97$$

How large should a total sample size be if we decide

(a) To pre-set $\alpha = .05$?

(b) That it is important to detect an LDL cholesterol level reduction of 1.0 with a power of 90%?

7. A study will be conducted to determine if some literature on smoking will improve patient comprehension. All subjects will be administered a pre-test, then randomized into two groups: without or with a booklet. After a week, all subjects will be administered a second test. The data (differences between pre-score and post-score) for a pilot study ($n = 44$) without a booklet yielded

$$\bar{x} = .25, \qquad s = 2.28.$$

How large should a total sample size be if we decide

(a) To pre-set $\alpha = .05$?

(b) That it is important to detect a mean difference of 1.0 with power of 90%?

8. A study will be conducted to compare the proportions of unplanned pregnancy between condom users and pill users. Preliminary data show that these proportions are approximately 10% and 5%, respectively. How large should a total sample size be so that it would be able to detect a difference of 5% with a power of 90%?

9. Suppose we want to compare the use of medical care by black and white teenagers. The aim is to compare the proportions of kids without physical checkups within the last 2 years. A recent survey shows that these rates for blacks and whites are 17% and 7%, respectively. How large should a total sample be so that it would be able to detect such a 10% difference with a power of 90%?

10. The following data are taken from a study that compares adolescents who have bulimia with healthy adolescents with similar body compositions and levels of physical activity. The following table provides measures of daily caloric intake (kcal/kg) for random samples of 23 bulimic adolescents and 15 healthy ones.

Bulimic adolescents			Healthy adolescents	
15.9	17.0	18.9	30.6	40.8
16.0	17.6	19.6	25.7	37.4
16.5	28.7	21.5	25.3	37.1
18.9	28.0	24.1	24.5	30.6
18.4	25.6	23.6	20.7	33.2
18.1	25.2	22.9	22.4	33.7
30.9	25.1	21.6	23.1	36.6
29.2	24.5	23.8		

Use the Wilcoxon test to compare the population medians.

11. A group of 19 rats was randomly divided into two groups. The 12 animals in the experimental group lived together in a large cage, furnished with playthings that were changed daily, while the seven animals in the control group lived in isolation without toys. The following table provides the cortex weights (the thinking part of the brain) in milligrams:

Experimental group	Control group
707, 740, 745, 652, 649, 676, 699, 696, 712, 708, 749, 690	669, 650, 651, 627, 656, 642, 698

Use the Wilcoxon test to compare the population medians.

12. The following table compiles data from different studies designed to investigate the accuracy of death certificates. The results of 5,373 autopsies were compared

with the causes of death listed on the certificates. The results are:

Date of study	Accurate certificate		
	Yes	No	Total
1955–1965	2,040	694	2,734
1970–1971	437	203	640
1975–1978	1,128	599	1,727
1980	121	151	272

Test to confirm the downward trend of accuracy over time.

13. A study was conducted to ascertain factors that influence a physician's decision to transfuse a patient. A sample of 49 attending physicians was selected. Each physician was asked a question concerning the frequency with which an unnecessary transfusion was give because another physician suggested it. The same question was asked of a sample of 71 residents. The data were as follows:

Type of physician	Frequency of unnecessary transfusion				
	Very frequent (1/week)	Frequent (1/2 weeks)	Occasionally (1/month)	Rarely (1/2 months)	Never
Attending	1	1	3	31	13
Resident	2	13	28	23	5

Test the null hypothesis of no association, with attention to the natural ordering of the columns.

14. Consider the following observations on the attention span of a child (y, in minutes) and the child's IQ (x):

x	y
75	2.0
80	3.0
85	4.5
90	5.0
95	5.2
100	5.5
110	6.7
115	6.8
130	3.8
135	2.9
140	2.0

Calculate both the Spearman's rho and the Kendall's tau. (Any comments on the linearity?)

15. The following data are the rates of oxygen consumption (y) of eight birds, measured at different temperatures (x):

x (°C)	y (ml/g/hr)
−19	5.2
−15	4.7
−10	4.5
−5	3.4
0	3.6
5	3.1
10	2.7
19	1.8

Calculate the two rank correlation coefficients, Spearman's rho and Kendall's tau.

Concluding Remarks

WHERE SHOULD YOU GO FROM HERE?

Let's say you've done reasonably well in this course, understanding the subjects well and mastering most of the techniques. Then comes the question of taking another course.

One thing another course would do for you is to help fix the ideas of this course a little better in your mind. Statistics is not an easy subject, once you get past the stage of drilling in the elementary calculations, and it takes a lot of time for the ideas to soak in and become a part of you. There are many more advanced calculations available, but there are also many key principles underlying the elementary ones. Your main decision now is whether to focus on learning more advanced calculations or to focus on understanding basic principles. This decision rests on what is available in the way of courses and your own inclinations.

Commonly used calculations include more advanced topics in the analysis of variance, regression analysis, and analysis of covariance. Also popular are survival analysis formulas, categorical data analysis, and modern nonparametric statistics. Most large universities have courses covering all these topics from a "calculation" perspective. The rapid advancement in both personal computers and mainframe computers provides ever increasing ease in computations, but there is a strong case to be made for knowing which formulas the computers are using for the calculations.

We want to encourage, for a variety of reasons, the deepest possible understanding of the principles of statistical inference. Only with the understanding of these principles will you ever be able to move beyond the calculation stage into the study design stage. Statistical principles are not easy to learn, but once mastered are worth their weight in gold. Learning the concept of "power" of tests and the real meaning of confidence intervals takes time and hard study, but is essential to the interpretation of calculations. It is also vital to understand what is meant by a "random sample" and its importance to statistical application. If we were to pick out one concept that is crucial to understanding statistics, it would be the concept of standard error. *Standard error* sounds temptingly like *standard deviation*, but knowing the meanings of those two phrases, and the ways in which they are alike as well as the ways in which they are different, takes you to a higher level of statistical understanding and opens the door to advanced areas. The more you learn about statistics the more you will appreciate the importance of standard errors and their role in the Central Limit Theorem, the jewel in the crown of both statistical theory and statistical applications.

Important sources of statistics courses are, of course, biostatistics and statistics departments. To a mathematics department, statistics is simply one of many applications of mathematics. To a psychology or education department, statistics is a useful tool but no more than that. To a biostatistics or statistics department, the subject is its reason for existing. Taking the course from a statistics or biostatistics department has several obvious advantages, including concern and respect for the subject as well as plenty of expertise. The student shopping for a statistics course is well advised to look at the biostatistics or statistics department's offerings. There are two things to watch out for, however. First, some statistics departments are very mathematical even in their service courses and thus have the same shortcomings as courses in mathematics departments. Second, those that are not too mathematical use examples from a wide variety of fields. If you are an engineer, for example, you may have trouble getting interested in examples from psychology and yet find that the teacher's examples from engineering demonstrate his or her great ignorance of engineering. The teacher's ignorance of engineering may so irritate the engineering student that he or she cannot concentrate on learning the statistical principles, and may perhaps wonder whether there *are* applications of statistics in engineering.

From a statistician's point of view, all the elementary statistics courses that do not have calculus as a prerequisite are essentially alike; only the examples used make them different. The students from engineering want engineering examples only, while students from psychology want psychology examples only. There certainly are differences in the emphases of use in different fields. For example, engineers make wide use of "control charts," while psychologists emphasize a technique called *the analysis of variance*. Still, the underlying statistical principles are the same. The elementary courses are commonly called *service courses*. They are intended for students who will take it as their once-in-a-lifetime statistical experience. Only infrequently does the university offer a second service course naturally following the first service course.

There are also service courses *with* calculus as a prerequisite. The calculus prerequisite is used more as a screen for mathematical facility than as a body of knowledge needed for the material. In shopping for a course, find out whether calculus is a prerequisite.

Our favorite advice to anyone trying to learn statistics is to talk about it to as many other students as you can and read *several* books. The books for beginners in statistics all say pretty much the same thing, but one book may say it better for you than another book. The point in talking to other people is that to attempt to explain it will help clarify the subject to you. It doesn't matter which one of you is the better student. Even if the two students are at extreme ends of the spectrum of knowledge and ability, they both increase their understanding by discussing it. The better student will benefit by trying to clarify the concepts for the other person. We, after many years of teaching statistics, continue to deepen our own understanding of the subject through teaching elementary courses.

Many elementary statistics texts have several exercises at the ends of each chapter with answers to those exercises in the back of the book. Students should not bother doing an exercise (unless it is part of the basis for grading) if they can see exactly how to do it, and see at the outset that doing it would only be tedious and provide no learning experiences. If, however, they are not sure how to tackle it, they should try. Try to see beyond each particular exercise, and try to generalize the exercise's results from biology, say, to economics.

DOING YOUR OWN STATISTICS WITH STATISTICAL PACKAGES

There are things called statistical packages (SPs) like "recipes" for analysis of data, on computers; even smaller personal computers have statistical packages. Almost all of the statistical formulas are on the SPs of computers. As a result, many non-statisticians are not seeking the advice of a statistician, but are using their personal computer SPs or going to computer experts for statistical help. Are they behaving wisely?

The SPs on computers can perform the calculations for 90% of all statistical analysis. Not every package contains all the techniques, but there is probably a package somewhere that will do what you want done. The statistical issue is deciding which one you should use.

The availability of SPs to the public is analogous to the situation that would exist if all medicines and prescription drugs were sold over the counter. If physicians were not able to control the public's access to the medicines and drugs that the medical profession considers dangerous, and didn't even *inform* the public as to which are considered dangerous, health care would be drastically different. People would purchase and use antibiotics, steroids, and chemotherapies at will. The situation of SPs is analogous, in that non-statisticians are free to analyze their data as they wish, trying a sequence of analyses until they achieve desired results. Sometimes, we heartily agree with the analyses they choose and other times are appalled at the most inappropriate tortures to which they submit their data. It is frustrating to see them elicit advice from anyone who will listen to their analysis-decision quandary and then proceed to perform every analysis suggested by their casual advisers. They sometimes get lucky, with all the analyses leading to the same conclusion, and are thus not left with the problems of deciding after the fact which one(s) are appropriate. If the analyses lead to different conclusions, they are then in the awkward position of having to decide which analyses are appropriate, while trying to prove that they are not simply selecting *a posteriori* those that confirm their prejudices.

Generally speaking, those SPs that simply print out the data in tables and graphs are not leading to confusion and abuse. The disasters occur through the use of complicated statistical routines whose titles and descriptions beckon the unwary with promise of deep analysis at no risk to the investigator. The Analysis of Variance, for example, is a phrase that suggests the ability to take something apart, scientifically, and break it down to its smallest components. Factor Analysis attracts many followers; it suggests that it is able to discover essential groups of common threads. We frequently hear, "Let's do a factor analysis and see what falls out." One of the most used and useful SPs is called *multiple regression*. It attempts to use many variables in combination to explain yet another variable. For example, multiple regression is used to see how the number of auto fatalities varies with both speed laws and gasoline prices. As useful as multiple regression is, however, it is commonly seriously misinterpreted by those who don't understand the effects of interrelationships among the variables. It is a high-risk procedure when performed without statistical advice.

Bibliography: Index of Data Sets

Chapter 1

Examples

1.1. Minneapolis *Star Tribune*, Feb. 9, 1990.

1.2. Blot W.J. et al. (1978). Lung cancer after employment in shipyards during World War II. *New England Journal of Medicine* 299:620–624.

1.3. Hiller R. and Kahn A.H. (1976). Blindness from glaucoma. *British Journal of Ophthalmology* 80:62–69.

1.4. May D. (1974). Error rates in cervical cytological screening tests. *British Journal of Cancer* 29:106–113.

1.5. Minneapolis *Star Tribune*, Feb. 11, 1990.

1.6. Centers for Disease Control (1990). *Summary of notifiable diseases: United States 1989. Morbidity and Mortality Weekly Report* Oct. 1990, vol. 38. *Monthly Vital Statistics Report* Nov. 1990, vol. 39.

1.7. Associated Press, Feb. 1990.

1.8. Umen A.J. and Le C.T. (1986). Prognostic factors, models, and related statistical problems in the survival of end-stage renal disease patients on hemodialysis. *Statistics in Medicine* 5:637–652.

1.9. National Center for Health Statistics. 1977, pp. 1–47.

1.10. Yen S., Hsieh C., and MacMahon B. (1982). Consumption of alcohol and tobacco and other risk factors for pancreatitis. *American Journal of Epidemiology* 116:407–414.

1.11. Fox A.J. and Collier P.F. (1976). Low mortality rates in industrial cohort studies due to selection for work and survival in the industry. *British Journal of Preventive and Social Medicine* 30:225–230.

Exercises

1. Coren S. (1989). Left-handedness and accident-related injury risk. *American Journal of Public Health* 79:1040–1041.

2. Rossignol A.M. (1989). Tea and premenstrual syndrome in the People's Republic of China. *American Journal of Public Health* 79:67–68.

3. Li D.K. et al. (1990). Prior condom use and the risk of tubal pregnancy. *American Journal of Public Health* 80:964–966.

4. Grady W.R. et al. (1986). Contraceptives failure in the United States: Estimates from the 1982 National Survey of Family Growth. *Family Planning Perspectives* 18:200–209.

5. Ockene J. (1990). The relationship of smoking cessation to coronary heart disease and lung cancer in the Multiple Risk Factor Intervention Trial. *American Journal of Public Health* 80:954–958.

6. Strogatz D. (1990). Use of medical care for chest pain differences between blacks and whites. *American Journal of Public Health* 80:290–293.

7. *Morbidity and Mortality Weekly Report*. Oct. 1990, vol. 38.

8. Makuc D. et al. (1989). National trends in the use of preventive health care by women. *American Journal of Public Health* 79:21–26.

9. Begg C. B. and McNeil B. (1988). Assessment of radiologic tests: Control of bias and other design considerations. *Radiology* 167:565–569.

10. Khabbaz R. et al. (1990). Epidemiologic assessment of screening tests for antibody to human T lymphotropic virus type I. *American Journal of Public Health* 80:190–192.

15. Kleinman J.C. and Kopstein A. (1981). Who is being screened for cervical cancer? *American Journal of Public Health* 71:73–76.

16. Hollows F.C. and Graham P.A. (1966). Intraocular pressure, glaucoma, and glaucoma suspects in a defined population. *British Journal of Ophthalmology* 50:570–586.

21. Knowler W.C. et al. (1981). Diabetes incidence in Pima Indians: Contributions of obesity and parental diabetes. *American Journal of Epidemiology* 113:144–156.

22. Rosenberg L. et al. (1981). Case–control studies on the acute effects of coffee upon the risk of myocardial infarction: Problems in the selection of a hospital control series. *American Journal of Epidemiology* 113:646–652.

23. Dienstag J.L. and Ryan D.M. (1982). Occupational exposure to hepatitis B virus in hospital personnel: Infection or immunization. *American Journal of Epidemiology* 15:26–39.

24. Weinberg G.B. et al. (1982). The relationship between the geographic distribution of lung cancer incidence and cigarette smoking in Allegheny County, Pennsylvania. *American Journal of Epidemiology* 115:40–58.

25. Berkowitz G.S. (1981). An epidemiologic study of pre-term delivery. *American Journal of Epidemiology* 113:81–92.

26. Arsenault P.S. (1980). Maternal and antenatal factors in the risk of sudden infant death syndrome. *American Journal of Epidemiology* 111:279–284.

27. Strader C.H. et al. (1988). Vasectomy and the incidence of testicular cancer. *American Journal of Epidemiology* 128:56–63.

28. True W.R. et al. (1988). Stress symptomology among Vietnam veterans. *American Journal of Epidemiology* 128:85–92.

29. Negri E. et al. (1988). Risk factors for breast cancer: Pooled results from three Italian case-control studies. *American Journal of Epidemiology* 128:1207–1215.

30. Graham S. et al. (1988). Dietary epidemiology of cancer of the colon in western New York. *American Journal of Epidemiology* 128:490–503.

Chapter 2

Examples

2.1. National Center for Health Statistics. 1987, p. 112.

2.4. Einarsson K. et al. (1985). Influence of age on secretion of cholesterol and synthesis of bile acids by the liver. *New England Journal of Medicine* 313:277–282.

2.7. Freireich E.J. et al. (1963). The effect of 6-mercaptopurine on the duration of steroid-induced remissions in acute leukemia: A model for evaluation of other potentially useful therapy. *Blood* 21:699–716.

Exercises

1. Fulwood R. et al. (1986). Total serum cholesterol levels of adults 20–74 years of age: United States, 1976–1980. *Vital and Health Statistics, Series M*, No. 236.

2. Yassi A. et al. (1991). An analysis of occupational blood level trends in Manitoba: 1979 through 1987. *American Journal of Public Health* 81:736–740.

3. Erdelyi G.J. (1976). Effects of exercise on the menstrual cycle. *The Physician and Sports Medicine*. Mar. 1976, p. 79.

11. Koenig J.Q. et al. (1990). Prior exposure to ozone potentiates subsequent response to sulfur dioxide in adolescent asthmatic subjects. *American Review of Respiratory Disease* 141:377–380.

12. Sandek C.D. et al. (1989). A preliminary trial of the programmable implantable medication system for insulin delivery. *The New England Journal of Medicine* 321:574–579.

13. *American Forests*. Apr. 1990, p. 71.

14. Gwirtsman H.E. et al. (1989). Decreased caloric intake in normal weight patients with bulimia: Comparison with female volunteers. *American Journal of Clinical Nutrition* 49:86–92.

15. Douglas G. (1990). Drug therapy. *The New England Journal of Medicine* 322: 443–449.

Chapter 3

Examples
3.7. *Journal of the American Medical Association*, 1964.

Exercises
 2. Klinhamer P.J.J.M. et al.(1989). Intraobserver and interobserver variability in the quality assessment of cervical smears. *Acta Cytologica* 33:215–218.
14. Biracree T. (1984). Your intelligence quotient. In *How You Rate*. New York: Dell Publishing Co., Inc.

Chapter 4

Exercises
 6. Koenig J.Q., Covert D.S., Hanley Q.S. et al. (1990). Prior exposure to ozone potentiates subsequent response to sulfur dioxide in adolescent asthmatic subjects. *American Review of Respiratory Disease* 141:377–380.
 7. Saudek C.D., Selam J.L., Pitt, H.A. et al. (1989). A preliminary trial of the programmable implantable medication system for insulin delivery. *The New England Journal of Medicine* 321:574–579.
13. Engs R.C. and Hanson D.J. (1988). University students' drinking patterns and problems: Examining the effects of raising the purchase age. *Public Health Reports* 103:667–673.
14. Nurminen M. et al. (1982). Quantitated effects of carbon disulfide exposure, elevated blood pressure and aging on coronary mortality. *American Journal of Epidemiology* 115:107–118.
15. Whittemore A.S. et al. (1988). Personal and environmental characteristics related to epithelial ovarian cancer. *American Journal of Epidemiology* 128:1228–1240.
16. Pappas G. et al. (1990). Hypertension prevalence and the status of awareness, treatment, and control in the Hispanic health and nutrition examination survey (HHANES). *American Journal of Public Health* 80:1431–1436.
17. Renaud L. and Suissa S. (1989). Evaluation of the efficacy of simulation games in traffic safety education of kindergarten children. *American Journal of Public Health* 79:307–309.
18. Fiskens E.J.M. and Kronshout D. (1989). Cardiovascular risk factors and the 25 year incidence of diabetes mellitus in middle-aged men. *American Journal of Epidemiology* 130:1101–1108.

19. Matinez F.D. et al. (1992). Maternal age as a risk factor for wheezing lower respiratory illness in the first year of life. *American Journal of Epidemiology* 136:1258–1268.

20. Whittemore A.S. et al. (1992). Characteristics relating to ovarian cancer risk: Collaborative analysis of 12 U.S. case–control studies. *American Journal of Epidemiology* 136:1184–1203.

Chapter 6

Examples

6.2. Rousch G.C. et al. (1982). Scrotal carcinoma in Connecticut metal workers: Sequel to a study of sinonasal cancer. *American Journal of Epidemiology* 116: 76–85.

6.3. Helsing K.J. and Szklo M. (1981) Mortality after bereavement. *American Journal of Epidemiology* 114:41–52.

6.6. Blot W.J. et al. (1978) Lung cancer after employment in shipyards during World War II. *The New England Journal of Medicine* 299:620–624.

6.7. Shapiro S. et al. (1979) Oral-contraceptive use in relation to myocardial infarction. *Lancet* 1:743–747.

6.8. Jekel J.F., Freeman D.H., and Mergs J.W. (1978) A study of trends in upper arm soft tissue sarcomas in the State of Connecticut following the introduction of alum-adsorbed allergic extract. *Annals of Allergy* 40:28–31.

6.13. Taylor H. (1981). Racial variations in vision. *American Journal of Epidemiology* 113:62–80.

Exercises

2. Mack T.M. et al. (1976). Estrogens and endometrial cancer in a retirement community. *New England Journal of Medicine* 294:1262–1267.

3. Padian N.S. (1990). Sexual histories of heterosexual couples with one HIV-infected partner. *American Journal of Public Health* 80:990–991.
   Helsing K.J. and Szklo M. (1981). Mortality after bereavement. *American Journal of Epidemiology* 114:41–52.

4. Schwarts B. et al. (1989). Olfactory function in chemical workers exposed to acrylate and methacrylate vapors. *American Journal of Public Health* 79:613–618.

5. D'Angelo L.J. et al. (1981). Epidemic keratoconjunctivitis caused by adenovirus Type 8: Epidemiologic and laboratory aspects of a large outbreak. *American Journal of Epidemiology* 113:44–49.

6. Reeves R. et al. (1981). Transmission of multiple drug-resistant tuberculosis: Report of a school and community outbreaks. *American Journal of Epidemiology* 113:423–435.

7. Shapiro S. et al. (1979). Oral contraceptive use in relation to myocardial infarction. *Lancet* 1:743–746.
   McCusker J. et al. (1988). Association of electronic fetal monitoring during labor with caesarean section rate with neonatal morbidity and mortality. *American Journal of Public Health* 78:1170–1174.

8. Thompson R.S. et al. (1989). A case–control study of the effectiveness of bicycle safety helmets. *The New England Journal of Medicine* 320:1361–1367.

9. Tuyns A.J. et al. (1977). Esophageal cancer in Ille-et-Vilaine in relation to alcohol and tobacco cnsumption: Multiplicative risks. *Bulletin of Cancer* 64:45–60.

11. Kono S. et al. (1992). Prevalnce of gallstone disease in relation to smoking, alcohol use, obesity, and glucose tolerance: A study of self-defense officials in Japan. *American Journal of Epidemiology* 136:787–805.

12. Jackson R. et al. (1992). Does recent alcohol consumption reduce the risk of acute myocardial infarction and coronary death in regular drinkers? *American Journal of Epidemiology* 136:819–824.

14. Kelsey J.L. et al. (1982). A case–control study of cancer of the endometrium. *American Journal of Epidemiology* 116:333–342.

15. Nischan P. et al. (1988). Smoking and invasive cervical cancer risk: Results from a case–control study. *American Journal of Epidemiology* 128:74–77.

18. Anderson J.W. et al. (1990). Oat bran cereal lowers serum cholesterol total and LDL cholesterol in hypercholesterolemic men. *American Journal of Clinical Nutrition* 52:495–499.

22. Frericks R.R. et al. (1981). Prevalence of depression in Los Angeles County. *American Journal of Epidemiology* 113:691–699.

23. Palta M. et al. (1982). Comparison of self-reported and measured height and weight. *American Journal of Epidemiology* 115:223–230.

24. Selby J.V. et al. (1989). Precursors of essential hypertension: The role of body fat distribution. *American Journal of Epidemiology* 129:43–53.

25. Taylor P.R. et al. (1989). The relationship of polychlorinated biphenyls to birth weight and gestational age in the offspring of occupationally exposed mothers. *American Journal of Epidemiology* 129:395–406.

26. Lee M. (1989). Improving patient comprehension of literature on smoking. *American Journal of Public Health* 79:1411–1412.

27. Fowkes F.G.R. et al. (1992). Smoking, lipids, glucose intolerance, and blood pressure as risk factors for peripheral atherosclerosis compared with ischemic heart disease in the Edinburgh Artery Study. *American Journal of Epidemiology* 135:331–340.

30. Kushi L.H. et al. (1988). The association of dietary fat with serum cholesterol in vegetarians: The effects of dietary assessment on the correlation coefficient. *American Journal of Epidemiology* 128:1054–1064.

Chapter 7

Exercises

10. Gwirtsman H.E. et al. (1989). Decreased caloric intake in normal-weight patients with bulimia: Comparison with female volunteers. *American Journal of Clinical Nutrition* 49:86–92.

12. Kircher T., Nelson J., and Burdo H. (1985) The autopsy as a measure of accuracy of death certificate. *The New England Journal of Medicine* 313:1263–1269.

13. Salem-Schatz S. et al. (1990). Influence of clinical knowledge, organization context and practice style on transfusion decision making. *Journal of the American Medical Association* 25:476–483.

Appendices

APPENDIX A
TABLE OF RANDOM NUMBERS

63271	59986	71744	51102	15141	80714	58683	93108	13554	79945
88547	09896	95436	79115	08303	01041	20030	63754	08459	28364
55957	57243	83865	09911	19761	66535	40102	26646	60147	15704
46276	87453	44790	67122	45573	84358	21625	16999	13385	22782
55363	07449	34835	15290	76616	67191	12777	21861	68689	03263
69393	92785	49902	58447	42048	30378	87618	26933	40640	16281
13186	29431	88190	04588	38733	81290	89541	70290	40113	08243
17726	28652	56836	78351	47327	18518	92222	55201	27340	10493
36520	64465	05550	30157	82242	29520	69753	72602	23756	54935
81628	36100	39254	56835	37636	02421	98063	89641	64953	99337
84649	48968	75215	75498	49539	74240	03466	49292	36401	45525
63291	11618	12613	75055	43915	26488	41116	64531	56827	30825
70502	53225	03655	05915	37140	57051	48393	91322	25653	06543
06426	24771	59935	49801	11082	66762	94477	02494	88215	27191
20711	55609	29430	70165	45406	78484	31639	52009	18873	96927

41990	70538	77191	25860	55204	73417	83920	69468	74972	38712
72452	36618	76298	26678	89334	33938	95567	29380	75906	91807
37042	40318	57099	10528	09925	89773	41335	96244	29002	46453
53766	52875	15987	46962	67342	77592	57651	95508	80033	69828
90585	58955	53122	16025	84299	53310	67380	84249	25348	04332
32001	96293	37203	64516	51530	37069	40261	61374	05815	06714
62606	64324	46354	72157	67248	20135	49804	09226	64419	29457
10078	28073	85389	50324	14500	15562	64165	06125	71353	77669
91561	46145	24177	15294	10061	98124	75732	00815	83452	97355
13091	98112	53959	79607	52244	63303	10413	63839	74762	50289
73864	83014	72457	22682	03033	61714	88173	90835	00634	85169
66668	25467	48894	51043	02365	91726	09365	63167	95264	45643
84745	41042	29493	01836	09044	51926	43630	63470	76508	14194
48068	26805	94595	47907	13357	38412	33318	26098	82782	42851
54310	96175	97594	88616	42035	38093	36745	56702	40644	83514
14877	33095	10924	58013	61439	21882	42059	24177	58739	60170
78295	23179	02771	43464	59061	71411	05697	67194	30495	21157
67524	02865	39593	54278	04237	92441	26602	63835	38032	94770
58268	57219	68124	73455	83236	08710	04284	55005	84171	42596
97158	28672	50685	01181	24262	19427	52106	34308	73685	74246
04230	16831	69085	30802	65559	09205	71829	06489	85650	38707
94879	56606	30401	02602	57658	70091	54986	41394	60437	03195
71446	15232	66715	26385	91518	70566	02888	79941	39684	54315
32886	05644	79316	09819	00813	88407	17461	73925	53037	91904
62048	33711	25290	21526	02223	75947	66466	06232	10913	75336
84534	42351	21628	53669	81352	95152	08107	98814	72743	12849
84707	15885	84710	35866	06446	86311	32648	88141	73902	69981
19409	40868	64220	80861	13860	68493	52908	26374	63297	45052
57978	48015	25973	66777	45924	56144	24742	96702	88200	66162
57295	98298	11199	96510	75228	41600	47192	43267	35973	23152
94044	83785	93388	07833	38216	31413	70555	03023	54147	06647
30014	25879	71763	96679	90603	99396	74557	74224	18211	91637
07265	69563	64268	88802	72264	66540	01782	08396	19251	83613
84404	88642	30263	80310	11522	57810	27627	78376	36240	48952
21778	02085	27762	46097	43324	34354	09369	14966	10158	76089

APPENDIX B
AREAS UNDER THE STANDARD NORMAL CURVE

z	.00	.01	.02	.03	.04	.05	.06	.07	.08	.09
.0	.0000	.0040	.0080	.0120	.0160	.0199	.0239	.0279	.0319	.0359
.1	.0398	.0438	.0478	.0517	.0557	.0596	.0636	.0675	.0714	.0753
.2	.0793	.0832	.0871	.0910	.0948	.0987	.1026	.1064	.1103	.1141
.3	.1179	.1217	.1255	.1293	.1331	.1368	.1406	.1443	.1480	.1517
.4	.1554	.1591	.1628	.1664	.1700	.1736	.1772	.1808	.1844	.1879
.5	.1915	.1950	.1985	.2019	.2054	.2088	.2123	.2157	.2190	.2224
.6	.2257	.2291	.2324	.2357	.2389	.2422	.2454	.2486	.2518	.2549
.7	.2580	.2612	.2642	.2673	.2704	.2734	.2764	.2794	.2823	.2852
.8	.2881	.2910	.2939	.2967	.2995	.3023	.3051	.3078	.3106	.3133
.9	.3159	.3186	.3212	.3238	.3264	.3289	.3315	.3340	.3365	.3389
1.0	.3413	.3438	.3461	.3485	.3508	.3531	.3665	.3577	.3599	.3621
1.1	.3643	.3554	.3686	.3708	.3729	.3749	.3770	.3790	.3810	.3830
1.2	.3849	.3869	.3888	.3907	.3925	.3944	.3962	.3980	.3997	.4015
1.3	.4032	.4049	.4066	.4082	.4099	.4115	.4131	.4147	.4162	.4177
1.4	.4192	.4207	.4222	.4236	.4251	.4265	.4279	.4292	.4306	.4319
1.5	.4332	.4345	.4357	.4370	.4382	.4394	.4406	.4418	.4429	.4441
1.6	.4452	.4463	.4474	.4484	.4495	.4505	.4515	.4525	.4535	.4545
1.7	.4554	.4564	.4573	.4582	.4591	.4599	.4608	.4616	.4625	.4633
1.8	.4641	.4649	.4656	.4664	.4671	.4678	.4686	.4693	.4699	.4706
1.9	.4713	.4719	.4726	.4732	.4738	.4744	.4750	.4756	.4761	.4767
2.0	.4772	.4778	.4783	.4788	.4793	.4798	.4803	.4808	.4812	.4817
2.1	.4821	.4826	.4830	.4834	.4838	.4842	.4846	.4850	.4854	.4857
2.2	.4861	.4864	.4868	.4871	.4875	.4878	.4881	.4884	.4887	.4890
2.3	.4893	.4896	.4898	.4901	.4904	.4906	.4909	.4911	.4913	.4916
2.4	.4918	.4920	.4922	.4925	.4927	.4929	.4931	.4932	.4934	.4936
2.5	.4938	.4940	.4941	.4943	.4945	.4946	.4948	.4949	.4951	.4942
2.6	.4953	.4955	.4956	.4957	.4959	.4960	.4961	.4962	.4963	.4964
2.7	.4965	.4966	.4967	.4968	.4969	.4970	.4971	.4972	.4973	.4974
2.8	.4974	.4975	.4976	.4977	.4977	.4978	.4979	.4979	.4980	.4981
2.9	.4981	.4982	.4982	.4983	.4984	.4984	.4985	.4985	.4986	.4986
3.0	.4986	.4987	.4987	.4988	.4988	.4989	.4989	.4989	.4990	.4990

Entries in the table give the area under the curve between the mean and z standard deviations above the mean. For example, for $z = 1.25$ the area under the curve between the mean and z is .3944.

APPENDIX C
PERCENTILES OF THE *t*-DISTRIBUTION

Degrees of freedom	Area in upper tail				
	.10	.05	.025	.01	.005
1	3.078	6.314	12.706	31.821	63.657
2	1.886	2.920	4.303	6.965	9.925
3	1.638	2.353	3.182	4.541	5.841
4	1.533	2.132	2.776	3.747	4.604
5	1.476	2.015	2.571	3.365	4.032
6	1.440	1.943	2.447	3.143	3.707
7	1.415	1.895	2.365	2.998	3.499
8	1.397	1.860	2.306	2.896	3.355
9	1.383	1.833	2.262	2.821	3.250
10	1.372	1.812	2.228	2.764	3.169
11	1.363	1.796	2.201	2.718	3.106
12	1.356	1.782	2.179	2.681	3.055
13	1.350	1.771	2.160	2.650	3.012
14	1.345	1.761	2.145	2.624	2.977
15	1.341	1.753	2.131	2.602	2.947
16	1.337	1.746	2.120	2.583	2.921
17	1.333	1.740	2.110	2.567	2.898
18	1.330	1.734	2.101	2.552	2.878
19	1.328	1.729	2.093	2.539	2.861
20	1.325	1.725	2.086	2.528	2.845
21	1.323	1.721	2.080	2.518	2.831
22	1.321	1.717	2.074	2.508	2.819
23	1.319	1.714	2.069	2.500	2.807
24	1.318	1.711	2.064	2.492	2.797
25	1.316	1.708	2.060	2.485	2.787
26	1.315	1.706	2.056	2.479	2.779
27	1.314	1.703	2.052	2.473	2.771
28	1.313	1.701	2.048	2.467	2.763
29	1.311	1.699	2.045	2.462	2.756
30	1.310	1.697	2.042	2.457	2.750
40	1.303	1.684	2.021	2.423	2.704
60	1.296	1.671	2.000	2.390	2.660
120	1.289	1.658	1.980	2.358	2.617
α	1.282	1.645	1.960	2.326	2.576

Entries in the table give t_α values, where α is the area or probability in the upper tail of the *t*-distribution. For example, with 10 degrees of freedom and a .05 area in the upper tail, $t_{.05} = 1.812$.

Reprinted by permission of Biometrika Trustees from Table 12, Percentage Points of the *t*-Distribution, E.S. Pearson and H.O. Hartley, *Biometrika Tables for Statisticians*, Vol. I, 3rd Edition, 1966.

APPENDIX D
PERCENTILES OF THE CHI-SQUARE DISTRIBUTION

Degrees of freedom	Area in upper tail	
	.05	.01
1	3.84146	6.63490
2	5.99147	9.21034
3	7.81473	11.3499
4	9.48773	13.2767
5	11.0705	15.0863
6	12.5916	16.8119
7	14.0671	18.4753
8	15.5073	20.0902
9	16.9190	21.6660
10	18.3070	23.2093
11	19.6751	24.7250
12	21.0261	26.2170
13	22.3621	27.6883
14	23.6848	29.1413
15	24.9958	30.5779
16	26.2962	31.9999
17	27.5871	33.4087
18	28.8693	34.8053
19	30.1435	36.1908
20	31.4104	37.5662
21	32.6705	38.9321
22	33.9244	40.2894
23	35.1725	41.6384
24	36.4151	42.9798
25	37.6525	44.3141
26	38.8852	45.6417
27	40.1133	46.9630
28	41.3372	48.2782
29	42.5569	49.5879
30	43.7729	50.8922
40	55.7585	63.6907
50	67.5048	76.1539
60	79.0819	88.3794
70	90.5312	100.425
80	101.879	112.329
90	113.145	124.116
100	124.342	135.807

Entries in the table give χ^2_α values, where α is the area or probability in the upper tail of the chi-square distribution. For example, with 10 degrees of freedom and a .01 area in the upper tail, $\chi^2_{.01} = 23.2093$.

Reprinted by permission of Biometrika Trustees from Table 8, Percentage Points of the χ^2 Distribution, by E.S. Pearson and H.O. Hartley, *Biometrika Tables for Statisticians*, Vol. I, 3rd Edition, 1966.

Answers to Selected Exercises

CHAPTER 1

1. For left-handed: $p = .517$
   For right-handed: $p = .361$

2. For factory workers: $x = 49$
   For nursing students: $x = 73$

3. For cases: $p = .775$
   For controls: $p = .724$

9. Sensitivity = .733
   Specificity = .972

10. For Dupont's EIA:
   Sensitivity = .938
   Specificity = .988
   For cellular product's EIA:
   Sensitivity = 1.0
   Specificity = .962

11. (a) Percent of total deaths:

Heart disease	30.2%
Cancer	24.1%
Cerebrovascular disease	8.2%
Accidents	4.0%
Other causes	33.4%

   (b) 1991 population for Minnesota: 3,525,297

   (c) Rates per 100,000 population:

Cancer	235.4
Cerebrovascular disease	80.3
Accidents	39.2
Other causes	325.5

13. Odds ratio = 1.43

231

14. Odds ratios

   (a) For non-smokers: 1.28
   (b) For smokers: 1.61
   (c) $1.28 \neq 1.61$

15. (a) Odds ratios associated with race (black vs. white)

   For 25–44 years + poor :   1.06
   For 45–64 years + non-poor :   2.00
   For 65 + years + non-poor :   1.52

   Different ratios indicate possible effect modifications by age and/or income.

   (b) Odds ratios associated with income (poor vs. non-poor)

   For 25–44 years + black :   2.46
   For 45–64 years + black :   1.64
   For 65 + years + black :   1.18

   Different ratios indicate a possible effect modification by age.

   (c) Odds ratios associated with race (black vs. white)

   For 65 + years + poor :   1.18
   For 65 + years + non-poor :   1.52

   The difference ($1.18 \neq 1.52$) indicates a possible effect modification by income.

16. Sensitivity = .650
    Specificity = .917

17. (a) For 1987:   24,027
        For 1986:   15,017

   (b) Number of cases of AIDS transmitted from mothers to newborns in 1988: 468

18. Follow-up death rates:

Age (years)	Deaths/1,000 months
21–30	3.95
31–40	5.05
41–50	11.72
51–60	9.80
61–70	10.19
70+	20.19

RR(70+ years vs. 51–60 years) = 2.05

19. Death rates for Georgia (per 100,000)

   Crude rate: 908.3
   Adjusted rate   1060.9
      (U.S. as standard):
   Adjusted rate   228.2
      (Alaska as standard):

20. Standardized mortality ratios

	Years since entering the industry			
	1–4	5–9	10–14	15+
SMR	.215	.702	.846	.907

RR(15+ years vs. 1–4 years): 4.22

23. (a)

Group	Proportion of HBV positive workers
Physicians	
Frequent	.210
Infrequent	.079
Nurses	
Frequent	.212
Infrequent	.087

(b) Odds ratios associated with frequent contacts

    For physicians: 3.11

    For nurses: 2.80

26. (a)

Group	SMR
Male	1.22
Female	.78
Black	2.09
White	.85

(b) Relative risks

    Associated with gender: 1.56

    Associated with race: 2.46

27. Odds ratios

    For Protestants: .50

    For Catholics: 4.69

    For others: .79

Evidence of an effect modification (4.69 ≠ .50, .79)

28.

Sympton	Odds ratio
Nightmares	3.72
Sleep problems	1.54
Troubled memories	3.46
Depression	1.46
Temper control problems	1.78
Life goal association	1.50
Omit feelings	1.39
Confusion	1.57

29.

Group	Odds ratio
Males	
Low fat, low fiber	1.15
High fat, high fiber	1.80
High fat, low fiber	1.81
Females	
Low fat, low fiber	1.72
High fat, high fiber	1.85
High fat, low fiber	2.20

CHAPTER 2

1. Median = 193
2. Medians:

    For 1979 group: 47.6

    For 1987 group: 27.8

7. Median from graph: 83
   Exact: 86

8.

	$\bar{x}$	s^2	s	$s/\bar{x}$
Men	84.71	573.68	23.95	28.3%
Women	88.51	760.90	27.58	31.2%

10.

$$\bar{x} = 168.70$$
$$s^2 = 1368.12$$
$$s = 36.99$$

11.

$$\bar{x} = 3.05$$
$$s^2 = .37$$
$$s = .61$$

12.

$$\overline{x} = 112.78$$

$$s^2 = 208.07$$

$$s = 14.42$$

13.

$$\overline{x} = 169.04$$

$$s^2 = 81.54$$

$$s = 9.03$$

$$\text{Median} = 166$$

$$s/\overline{x} = 5.34\%$$

14.

Adolescents	$\overline{x}$	s^2
Bulimic	22.08	20.91
Healthy	29.70	42.13

15.

Drug	$\overline{x}$	s^2
A	133.95	503.83
R	267.35	4,448.99

16. (a)

$$\overline{x} = 8.67$$

$$s^2 = 41.83$$

$$s = 6.47$$

(b) Median $= 8 < \overline{x}$; distribution is skewed to the right

(c) Not directly because of censoring. Use Kaplan-Meier curve: median = 22.5

17. (a) 3, 4$^+$, 5.7$^+$, 6.5, 6.5, 8.4$^+$, 10$^+$, 10, 12, 15

CHAPTER 3

1. Odds ratio = .62
   Pr(Pap test = yes) = .82
   Pr(Pap test = yes | black) = .75 $\neq$ .82
2. Odds ratio = 5.99
   Pr(second screening = present) = .65
   Pr(second screening = present | first screening = present) = .78 $\neq$ .65
3. (a) .202
   (b) .217
   (c) .376
   (d) .268
4. Positive predictive value

   For Population A: .991
   For Population B: .900

5. (a) Sensitivity = .733, specificity = .972
   (b) .016
   (c) .301
6.

Prevalence	Positive predictive value
.2	.867
.4	.946
.6	.975
.7	.984
.8	.991
.9	.996

Yes:

$$\text{Prevalence}$$
$$= \frac{(\text{PPV})(1 - \text{specificity})}{(\text{PPV})(1 - \text{specificity}) + (1 - \text{PPV})(\text{sensitivity})}$$
$$= .133 \quad \text{if} \quad \text{PPV} = .8$$

7. (a) .1056
   (b) .7995
8. (a) .9500
   (b) .0790
   (c) .6992
9. (a) .9573
   (b) .1056
10. (a) 1.645
    (b) 1.96
    (c) .84
11.

   20–24 years: 101.23 106.31 141.49 146.57
   25–29 years: 104.34 109.00 141.20 145.86

12. (a) 200+ days: .1587; 365+ days: $\cong 0$
    (b) .0228
13. (a) .0808
    (b) 84.2
14. (a) 17.2
    (b) 19.2%
    (c) .0409
15. (a) .5934
    (b) .0475
    (c) .0475
16. (a) $\cong 0$
    (b) $\cong 0$
17. (a) .2266
    (b) .0045
    (c) .0014
18. (a) .0985
    (b) .0019
19.

   Rate = 13.89 per 1000 live births
   $z = 13.52$

20. (a) For 2.086: .975; for 2.845: .995
    (b) For 1.725: .05; for 2.528: .01
    (c) ± 2.086: .05; ± 2.845: .01

CHAPTER 4

1. $\mu = .5$; 6 possible samples with $\mu_{\overline{x}} = .5$
   $(= \mu)$
2.

$$\Pr(\mu - 1 \le \overline{x} \le \mu + 1) = \Pr(-2.33 \le z \le 2.33)$$
$$= .9786$$

3.

$$\overline{x} = 88.51$$
$$s = 27.58$$
$$\text{SE}(\overline{x}) = 5.12$$
$$88.51 \pm (1.96)(5.12) = (78.47, 98.55)$$
$$88.51 \pm (2.048)(5.12) = (78.02, 99.00)$$

4.

$$\overline{x} = 36.70$$
$$s = 15.90$$
$$\text{SE}(\overline{x}) = 2.11$$
$$36.70 \pm (1.96)(2.11) = (32.57, 40.83)$$

5.

$$\overline{x} = 152.03$$
$$s = 48.53$$
$$\text{SE}(\overline{x}) = 19.81$$
$$152.03 \pm (2.571)(19.81) = (101.09, 202.97)$$

6.

$$\overline{x} = 3.05$$
$$s = .61$$
$$\text{SE}(\overline{x}) = .19$$
$$3.05 \pm (2.262)(.19) = (2.61, 3.49)$$

7.

$$\bar{x} = 112.78$$

$$s = 14.42$$

$$SE(\bar{x}) = 3.40$$

$$112.78 \pm (2.110)(3.40) = (104.83, 119.95)$$

8. Men:

$$\bar{x}_1 = 84.65$$

$$s_1^2 = 576.00$$

Women:

$$\bar{x}_2 = 88.51$$

$$s_2^2 = 760.66$$

$$SE(\bar{x}_2 - \bar{x}_1) = 6.69$$

$$(88.51 - 84.65) \pm (1.96)(6.69) = (-9.26, 16.98)$$

9.

$$\text{Specificity } p = .917$$

$$SE(p) = .004$$

$$.917 \pm (1.96)(.004) = (.909, .925)$$

10. (a)

$$p = .175$$

$$SE(p) = .017$$

$$.175 \pm (1.96)(.017) = (.141, .209)$$

(b)

$$\text{Odds ratio} = \frac{(893)(83)}{(132)(392)}$$

$$= 1.43$$

$$\text{Exp}\left[\ln 1.43 \pm 1.96\sqrt{\frac{1}{893} + \frac{1}{83} + \frac{1}{132} + \frac{1}{392}}\right]$$

$$= (1.06, 1.93)$$

11.

$$\text{Odds ratio} = \frac{(13)(9993)}{(4987)(7)}$$

$$= 3.72$$

$$\text{Exp}\left[\ln 3.72 \pm 1.96\sqrt{\frac{1}{13} + \frac{1}{9993} + \frac{1}{4987} + \frac{1}{7}}\right]$$

$$= (1.48, 9.33)$$

12.

$$\text{Controls: } p_1 = .627$$

$$\text{Cases: } p_2 = .283$$

$$SE(p_1 - p_2) = .070$$

$$(.627 - .283) \pm (1.96)(.070) = (.207, .481)$$

13.

$$\text{For 1983: } p_1 = .474$$

$$\text{For 1987: } p_2 = .373$$

$$SE(p_2 - p_1) = .014$$

$$(.373 - .474) \pm (1.96)(.014) = (-.127, -.075)$$

14.

DBP group	95% C.I.
< 95:	(−9.48, 7.48)
95–100:	(−26.0, 8.0)
≥ 100:	(−17.76, 51.76)

No indication of any differences.

15. (a) Cases vs. hospital controls:

$$\text{Odds ratio} = \frac{(177)(31)}{(11)(249)}$$

$$= 2.0$$

$$\text{Exp}\left[\ln 2.0 \pm 1.96\sqrt{\frac{1}{177} + \frac{1}{31} + \frac{1}{11} + \frac{1}{249}}\right]$$

$$= (.98, 4.09)$$

(b) Cases vs. population controls:

$$\text{Odds ratio} = \frac{(177)(26)}{(11)(233)}$$

$$= 1.78$$

$$\text{Exp}\left[\ln 1.78 \pm 1.96\sqrt{\frac{1}{177} + \frac{1}{26} + \frac{1}{11} + \frac{1}{233}}\right]$$

$$= (.86, 3.73)$$

16. For whites:
$25.3 \pm (1.96)(.9) = (23.54\%, 27.06\%)$
For blacks:
$38.6 \pm (1.96)(1.8) = (35.07\%, 42.13\%)$

17. Control:
$7.9 \pm (1.96)(3.7)/\sqrt{30} = (6.58, 9.22)$
Simulation game:
$10.1 \pm (1.96)(2.3)/\sqrt{33} = (9.32, 10.88)$

18. (a) $25.0 \pm (1.96)(2.7)/\sqrt{58} = (24.31, 25.69)$

(b) On the average, people with large body mass index are more likely to develop diabetes mellitus

19.

Maternal age (years)	OR (95% confidence interval)	
	Boys	**Girls**
< 21	2.32 (.90, 5.99)	1.62 (.62, 4.25)
21–25	2.24 (1.11, 4.54)	1.31 (.74, 2.30)
26–30	1.55 (.78, 3.07)	1.31 (.76, 2.29)

20. (a)

Duration (years)	OR (95% confidence interval)
2–9	.94 (.74, 1.20)
10–14	1.04 (.71, 1.53)
≥ 15	1.54 (1.17, 2.02)

(b)

Group	OR (95% confidence interval)
No drug use	1.13 (.83, 1.53)
Drug use	3.34 (1.59, 7.02)

CHAPTER 5

1. (a) $\mathcal{H}_0 : \mu = 30$; $\mathcal{H}_A : \mu > 30$
   (b) $\mathcal{H}_0 : \mu = 11.5$; $\mathcal{H}_A : \mu \neq 11.5$
   (c) Hypotheses are for population parameters
   (d) $\mathcal{H}_0 : \mu = 31.5$; $\mathcal{H}_A : \mu < 31.5$
   (e) $\mathcal{H}_0 : \mu = 16$; $\mathcal{H}_A : \mu \neq 16$
   (f) Same as (c)

2. $\mathcal{H}_0 : \mu = 74.5$

3. $\mathcal{H}_0 : \mu = 7,250$; $\mathcal{H}_A : \mu < 7,250$

4. $\mathcal{H}_0 : \mu = 38$; $\mathcal{H}_A : \mu > 38$

5. $\mathcal{H}_0 : \pi = .007$; $\mathcal{H}_A : \mu > .007$
   p-value $= \Pr(\geq 20 \text{ cases out of } 1,000 \mid \mathcal{H}_0)$

$$\mathcal{H}_0 : \text{mean} = (1,000)(.007) = 7$$

$$\text{variance} = (1,000)(.007)(.993) = (2.64)^2$$

$$z = \frac{20 - 7}{2.64}$$

$$= 4.92; \qquad p\text{-value} \cong 0$$

8. $\mathcal{H}_0 : \sigma = 20; \quad \mathcal{H}_A : \sigma \neq 20$

9. One-sided

10. Under $\mathcal{H}_0 : \pi = .25$; variance $= \pi(1-\pi)/$
    $100 = (.043)^2$

$$z = \frac{.18 - .25}{.043}$$

$$= -1.63; \quad \alpha = .052$$

Under $\mathcal{H}_A : \pi = .15$; variance $= \pi(1-\pi)/$
$100 = (.036)^2$

$$z = \frac{.18 - .15}{.036}$$

$$= .83; \quad \beta = .2033$$

The change makes α smaller and β larger.

11. Under $\mathcal{H}_0$:

$$z = \frac{.22 - .25}{.043}$$

$$= -.70; \quad \alpha = .242$$

Under $\mathcal{H}_A$:

$$z = \frac{.22 - .15}{.036}$$

$$= 1.94; \quad \beta = .026$$

The new change makes α larger and β smaller.

12.

$$p\text{-value} = \Pr(p < .18 \text{ or } p > .32)$$

$$= 2\Pr\left(z \geq \frac{.32 - .25}{.043} = 1.63\right)$$

$$= .1032$$

13.

$$p = .18$$

$$SE(p) = .036$$

$$.18 \pm (1.96)(.036) = (.109, .251)$$

which includes .25

14. $SE(\bar{x}) = .054$
    Cut point for $\alpha = .05$: $4.86 + (1.96)(.054) = 4.96$

$$\mu = 4.86 + .1$$

$$= 4.96$$

$$z = \frac{4.96 - 4.96}{.054}$$

$$= 0; \quad \text{power} = .5$$

15. $SE(\bar{x}) = 1.11$
    Cut points for $\alpha = .05$: $128.6 \pm$
    $(1.96)(1.11) = 126.4$ and 130.8
    $\mu = 135$

$$\text{Power} = 1 - \Pr(126.4 \leq \bar{x} \leq 130.8 \mid \mu = 135)$$

$$\cong 1.0$$

CHAPTER 6

1. $p = .227$
   $z = -1.67; \quad p\text{-value} = (2)(.0475) = .095$

2. $z = 5.65; \quad p\text{-value} \cong 0$

3. $\mathcal{H}_0$: consistent report, i.e., man and woman agree

$$z = (6 - 7)/\sqrt{6 + 7}$$

$$= -.28; \quad p\text{-value} = .7794$$

4. $\mathcal{H}_0$: no effects of acrylate and methacrylate vapors on olfactory function

$$z = (22 - 9)/\sqrt{22 + 9}$$

$$= 2.33; \qquad p\text{-value} = .0198$$

5. $\chi^2 = 29.29; \quad p < .01$
6. $\chi^2 = 44.49; \quad p < .01$
7. $\chi^2 = 37.95; \quad p < .01$
8. $\chi^2 = 28.26; \quad p < .01$
9. For 25.44 years:

$$a = 5$$

$$\frac{r_1 c_1}{n} = 1.27$$

$$\frac{r_1 r_2 c_1 c_2}{n^2(n-1)} = 1.08$$

For 45–64 years:

$$a = 67$$

$$\frac{r_1 c_1}{n} = 32.98$$

$$\frac{r_1 r_2 c_1 c_2}{n^2(n-1)} = 17.65$$

For 65+ years:

$$a = 24$$

$$\frac{r_1 c_1}{n} = 13.28$$

$$\frac{r_1 r_2 c_1 c_2}{n^2(n-1)} = 7.34$$

$$z = \frac{(5 - 1.27) + (67 - 32.98) + (24 - 13.28)}{\sqrt{1.08 + 17.65 + 7.34}}$$

$$= 9.49; \qquad p \cong 0$$

10. $\chi^2 = 11.15; p < .01$
11. For smoking: $\chi^2 = .93; p > .05$
    For alcohol: $\chi^2 = .26; p > .05$
    For body mass index: $\chi^2 = .87; p > .05$

12. (a) Men
    For myocardial infarction: $\chi^2 = .16$; $p > .05$

    Odds ratio = .94;

    95% confidence interval = (.69, 1.28)

    For coronary death: $\chi^2 = 8.47$; $p < .01$

    Odds ratio = .57;

    95% confidence interval = (.39, .83)

    (b) Women
    For myocardial infarction: $\chi^2 = 4.09$; $p < .05$

    Odds ratio = .55;

    95% confidence interval = (.30, .99)

    For coronary death: $\chi^2 = 2.62$; $p > .05$

    Odds ratio = .39;

    95% confidence interval = (.12, 1.26)

13.

$$z = (264 - 249)/\sqrt{264 + 249}$$

$$= .66; \qquad p = .5092$$

14. < 57 kg:

$$a = 20$$

$$\frac{r_1 c_1}{n} = 9.39$$

$$\frac{r_1 r_2 c_1 c_2}{n^2(n-1)} = 5.86$$

57–75 kg:

$$a = 37$$

$$\frac{r_1 c_1}{n} = 21.46$$

$$\frac{r_1 r_2 c_1 c_2}{n^2(n-1)} = 13.60$$

> 75 kg:

$$a = 9$$

$$\frac{r_1 c_1}{n} = 7.63$$

$$\frac{r_1 r_2 c_1 c_2}{n^2(n-1)} = 4.96$$

$$z = \frac{(20 - 9.39) + (37 - 21.46) + (9 - 7.63)}{\sqrt{5.86 + 13.60 + 4.96}}$$

$$= 5.57; \quad p \cong 0$$

15. Zero or one:

$$a = 12$$

$$\frac{r_1 c_1}{n} = 6.94$$

$$\frac{r_1 r_2 c_1 c_2}{n^2(n-1)} = 4.48$$

Two or more:

$$a = 96$$

$$\frac{r_1 c_1}{n} = 93.22$$

$$\frac{r_1 r_2 c_1 c_2}{n^2(n-1)} = 28.65$$

$$z = \frac{(12 - 6.94) + (96 - 93.22)}{\sqrt{4.48 + 28.65}}$$

$$= 1.36; \quad p = .1732$$

16.

$$SE(\bar{x}) = .5$$

$$t = (7.84 - 7)/(.5)$$

$$= 1.68, \ 15 \ df; \quad p > .10$$

17.

$$\bar{d} = 200$$

$$s_d = 397.2$$

$$SE(\bar{d}) = 150.1$$

$$t = (200 - 0)/(150 - 1)$$

$$= 1.33, \ 6 \ df; \quad p > .20$$

18. Corn flakes: $\bar{x}_1 = 4.44$, $s_1^2 = .97$
    Oat bran: $\bar{x}_2 = 4.08$; $s_2^2 = 1.06$

$$s_p = 1.02$$

$$t = (4.44 - 4.08)/1.02\sqrt{1/14 + 1/14}$$

$$= .937; \quad p > .20$$

19.

$$s_p = .175$$

$$t = 3.40; \quad p \cong 0$$

20.

$$s_p = 7.9$$

$$t = 4.40; \quad p \cong 0$$

21. Treatment: $\bar{x}_1 = 701.9$, $s_1 = 32.8$
    Control: $\bar{x}_2 = 656.1$, $s_2 = 22.5$

$$s_p = 29.6$$

$$t = 3.45, \ 17 \ df; \quad p < .01$$

22.

$$s_p = 9.3$$

$$t = 4.71; \quad p \cong 0$$

23. (a)

$$s_p = 1.47$$

$$t = 2.70; \quad p < .01$$

(b)

$$s_p = 1.52$$

$$t = 1.19; \quad p > .20$$

(c) ≤ High school:

$$s_p = 1.53$$
$$t = 6.53; \qquad p \cong 0$$

≥ College:

$$s_p = 1.35$$
$$t = 3.20; \qquad p \cong 0$$

24.

SBP: $t = 12.11; \quad p \cong 0$

DBP: $t = 10.95; \quad p \cong 0$

BMI: $t = 6.71; \quad p \cong 0$

25. Weight gain:

$$s_p = 14.4$$
$$t = 2.30; \qquad p < .05$$

Birth weight:

$$s_p = 471.1$$
$$t = 2.08; \qquad p < .05$$

Gestational age:

$$s_p = 15.3$$
$$t \cong 0; \qquad p \cong .5$$

26. 5th-grade level vs. no booklet:

$$s_p = 2.41$$
$$t = 2.57; \qquad p \cong .01$$

10th-grade level vs. no booklet:

$$s_p = 2.33$$
$$t = .7; \qquad p > .20$$

27.

Group 1 vs. group 4: $\quad t = 4.66; \quad p \cong 0$

Group 2 vs. group 4: $\quad t = 2.52; \quad p < .05$

Group 3 vs. group 4: $\quad t = 3.24; \quad p < .01$

28. $r = -.99$

29.

$$r = \frac{5,786}{\sqrt{(20,192)(3,380)}}$$
$$= .70$$

$$t = (.70)\sqrt{\frac{14}{1 - (.70)^2}}$$
$$= 3.67, \ 14 \ df; \qquad p < .01$$

30.

$$t = .46\sqrt{\frac{44}{1 - (.46)^2}}$$
$$= 3.44, \ 44 \ df; \qquad p < .01$$

CHAPTER 7

1. (a) With 99% confidence

$$n_* = \frac{(2.58)^2(.25)}{(.01)^2}$$
$$= 16,641$$

With 95% confidence

$$n_* = \frac{(1.96)^2(.25)}{(.01)^2}$$
$$= 9,604$$

(b) With 99% confidence

$$n = \frac{(2.58)^2(.08)(.92)}{(.01)^2}$$

$$= 4,900$$

With 95% confidence

$$n = \frac{(1.96)^2(.08)(.92)}{(.01)^2}$$

$$= 2,828$$

2. It depends on how accurate we want the estimate to be.

3.

$$n = \frac{(1.96)^2(2.5/4)^2}{(.5)^2}$$

$$= 7$$

4.

$$n = \frac{(1.96)^2(1)^2}{(.5)^2}$$

$$= 16$$

5.

$$s \doteq (4,000 - 500)/4$$

$$= 875$$

$$n = \frac{(1.96)^2(875)^2}{(100)^2}$$

$$= 295$$

6.

$$N = 4(1.96 + 1.65)^2 \frac{(.97)^2}{(1)^2}$$

$$= 50$$

7.

$$N = 4(1.96 + 1.65)^2 \frac{(2.28)^2}{(1)^2}$$

$$= 271$$

8.

$$\pi = .075$$

$$N = \frac{4\{(2)(1.96)(.075)(.925) + 1.65[(.05)(.95) + (.1)(.9)]\}^2}{(.1 - .05)^2}$$

$$= 399$$

9.

$$\pi = .12$$

$$N = \frac{4\{(2)(1.96)(.12)(.88) + (1.65)[(.17)(.83) + (.07)(.93)]\}^2}{(.17 - .07)^2}$$

$$= 228$$

10. Sum of ranks:

For bulimic adolescents: 337.5
For healthy adolescents: 403.5

$$\mu_H = \frac{15(15 + 23 + 1)}{2}$$

$$= 292.5$$

$$\sigma_H = \sqrt{\frac{(15)(23)(15 + 23 + 1)}{12}}$$

$$= 33.5$$

$$z = \frac{403.5 - 292.5}{33.5}$$

$$= 3.31; \qquad p \cong 0$$

11. Sum of ranks:

Experimental group: 151
Control group: 39

$$\mu_E = \frac{12(12 + 7 + 1)}{2}$$

$$= 120$$

$$\sigma_E = \sqrt{\frac{(12)(7)(12 + 7 + 1)}{12}}$$

$$= 11.8$$

$$z = \frac{151 - 120}{11.8}$$

$$= 262; \qquad p = .0088$$

12.

$$C = 2,442,198$$

$$D = 1,496,110$$

$$S = 946,088$$

$$\sigma_S = 95,706$$

$$z = 9.89; \qquad p \cong 0$$

13.

$$C = 364$$

$$D = 2238$$

$$S = -1874$$

$$\sigma_S = 372$$

$$z = -5.03; \qquad p \cong 0$$

14.

$$\rho = 1 - \frac{(6)(213.5)}{(11)(120)}$$

$$= .03$$

$$\tau = \frac{31 - 23}{(11)(10)/2}$$

$$= .14$$

It is not a linear relationship.

15.

$$\rho = 1 - \frac{(6)(166)}{(8)(63)}$$

$$= -.976$$

$$\tau = \frac{1 - 27}{(8)(7)/2}$$

$$= -.929$$

Index

About the Authors

Chap T. Le is a professor of biostatistics at the School of Public Health of the University of Minnesota. Since 1978, Dr. Le has taught biostatistics to undergraduate, graduate, and professional students, as well as having served the last few years as major chairperson and director of graduate studies of the Graduate Program of Biostatistics. He received the Leonard M. Schuman Excellence in Teaching Award presented by the School of Public Health. His research has focused on the analyses of survival and categorical data from clinical trials and lab studies, and his methodological interests include survival analysis, ordered alternatives, correlated binary data, and ROC curve. Dr. Le is the author of another textbook (*Fundamentals of Biostatistical Inference*) and is the author or coauthor of more than eighty research articles in major biostatistical and biomedical journals.

James R. Boen served as a professor in the Division of Biostatistics at the School of Public Health of the University of Minnesota where he taught introductory courses to a wide variety of health science professionals and directed the Biostatistical Consulting Laboratory. Dr. Boen is a general-practice biostatistician with extensive experience and academic interest in the process of consulting with biomedical researchers; his research activities spanned many areas of biomedical science. Dr. Boen is serving as Associate Dean for Academic Affairs of the School of Public Health.